TRANSFORMATIONAL INFORMATION

KEEPERS OF THE AKASHIC RECORDS

MESSAGES FROM THE
KEEPERS OF THE AKASHA
THE ASCENDED MASTERS
& THE ARC ANGELS

BY

BILL FOSS

THE AKASHIC RECORDS CENTER
INSTITUTE FOR HIGHER STUDIES

Acknowledgements:

I would like to thank everyone who has contributed to the workshops and classes their time, resources and personal support during the past years of teaching abroad and during the production and constant vigilance in writing, publishing and producing these books. Also to all the practitioners, teachers and holy masters who have changed the world for the better with their presence, service and teachings and have made a personal impact on my life and the lives of countless others with continued inspiration for creativity, service, and healing.

Note: The information and exercises contained within this book are not intended to replace psychological counselling or medical attention. If you feel the need for help, contact a qualified service professional in the appropriate field. Do not attempt to engage in the meditation, visualization, or energy exercises in this book while operating tools, heavy machinery or a motor vehicle.

Note: This book has cutting edge information that is repeated in several areas intentionally as a teaching style, while building on different aspects along the same themes.

ISBN-13: 978-1-7339358-4-5 (White Wizard Publishing)
ISBN-10: 1-7339358-4-3 "Keepers of the Akashic Records"

WHITE WIZARD
PUBLISHING

TABLE OF CONTENTS

INTRODUCTION VII

PREFACE - QUEST FOR THE AKASHIC RECORDS 1

1. KEEPERS OF THE AKASHIC RECORDS 13

2. DIFFERENT LEVELS AND LOCATIONS OF THE AKASHIC RECORDS 23

3. YOUR SOUL TAPESTRY, TREE OF LIFE & THE FATES 41

MEDITATION EXERCISE 1 - JOURNEY TO THE TAPESTRY OF SOULS 59

MEDITATION EXERCISE 2 - JOURNEY TO YOUR TREE OF LIFE 67

MEDITATION EXERCISE 3 - JOURNEY TO MEET THE FATES 75

4. PERSPECTIVES OF SPACE AND TIME REALITY 83

5. UNDERSTANDING REALITIES WITHIN THE SUPER UNIVERSAL 93

6. CONSCIOUSNESS, PERCEPTION & AWARENESS 107

7. DELEGATES FROM BEYOND THE VEIL 119

MEDITATION EXERCISE 4 - THE SACRED CIRCLE 141

8. THE REMEMBRANCE OF HOME 149

9. SEPARATION AND UNITY 163

MEDITATION EXERCISE 5 - MEET YOUR GUIDES IN THE AKASHIC RECORDS 173

10. SYNCHRONICITY & SERENDIPITY 181

11. QUANTUM ENERGY & THE ENTRAINMENT OF CONSCIOUSNESS 193

12. MIND EVOLUTION & THE ASCENSION OF MAN 203

MEDITATION EXERCISE 6 - REPROGRAMMING THE VAGUS NERVE 225

MEDITATION EXERCISE 7 - LETTING GO 228

MEDITATION EXERCISE 8 - BODY SCANNING FOR CLEARING 235

13. ANGELIC INFLUENCE ON THE EARTH PLANE 241

14. AKASHIC INFLUENCE OF THE CIVILIZATION OF MAN 255

EXERCISE 9 - JOURNEY INTO THE AKASHIC RECORDS 271

15. DIFFERENT VERSIONS OF SELF & THE REALITIES THEY OCCUPY 279

16. INTERACTIVE HEALING & MANIFESTING THROUGH YOUR BOOK OF LIFE 297

EXERCISE 10 - ACTIVATING YOUR AKASHIC SOUL VISION 309

17. CREATIVE PERCEPTIONS FOR UNDERSTANDING CHANGE. 317

GLOSSARY 329

PRODUCTS & SERVICES 364

Dedication

May this Book Serve You
In Your Search for
Greater Understanding
as You Look Into
Dreams, Visions and Intuitions.
With Eyes that can See
and Ears that can Hear,
For You, the Reader,
And to All the
Workshop Sponsors & Participants
the World Over, Thanks for Helping to Make
This Book a Reality,
I Hope It Serves You Well, Enjoy!

ARC

Introduction

Keepers of the Akashic Records is a collection of insights gathered through meditations, visions, epiphanies, streaming visions and moments of enlightenment. This book contains information that is foundational to all of your other work with energy, intuition, healing and manifesting. There are inner keys to making it all work and the ways to relate them to your life in a workable manner. While there are many different styles of gathering messages, this body of work is not intended to glamorize or emulate any other such styles of channelling.

Within the many different dimensions and realms of existence there are many beings who interact with each other and those of us here on this side listening in to receive messages and meanings for our own lives and for others. On this planet through our lifetimes, from this side of the veil, we have always felt the urge to practice consciously 'tapping in' to receive messages and energy from the spiritual and beings who are other worldly in nature.

Common or classic metaphysical styles of channelling

involve opening a perceived connection with a being on a different plane of existence and then starting to talk or write in a stream while allowing that personality whether angelic, ascended master, alien, the voice of Creator or a human between lifetimes to bring messages through as a form of mediumship.

The challenge with the practice of channelling is that it sometimes has a tendency to regurgitate general messages which can be incomplete or redundant. It also can often be filtered through the ego, giving a person all those things they want to hear. If this happens you must learn to go deeper and reach for clearer and higher messages. It takes intentional practice and a clear third eye and mind to be able to 'vibe in' and get genuine info.

Sometimes in being hopeful in connecting with higher dimensional beings it is also possible to get a 'wrong number' when making the call or a skewed message. So listen intently. You can always use the word 'Truth' before anyone's name that you are addressing on the other side. You can also use the question "Truth, who are you?" If you do not get an answer, continue and they will tell you. If they do not answer then you will know that the connection may not be of virtue.

In an age when accurate and inspirational messages are important we can continue to meet these intentions with practice of clarity. So look deeper into a more stylistically personalized experience of asking and listening in order to bring clearer and more distinct messages through.

This is my newest offering in my continued studies and in some ways the capstone of my collective works, which is their works, concerning the Akashic Records. All of the messages and information in the books is a blend of channelled mystical information and a practical approach to explanation, description and application of energy practices and Akashic access processes.

When I sat down to read one of my books aloud while making sound recordings for the audio books, and listened back to the recording I realized that I was not the only one participating in the writing. My first thought was 'I didn't write this.' I started to understand a deeper and even simpler level of the way that the information was creatively being 'channelled' through my conscious mind.

It has always been very apparent to me that I was having help in translating the ideas, concepts and dynamics of the Records concerning healing, the intuitive arts and the constructs of the inter-joined realities and their dynamics surrounding the higher realms.

As I wrote the table of contents, I realized that this was going to be a uniquely different and important book in comparison and addition to the rest of the collection.

They were guiding, directing and motivating me into new areas of thought, presence and reality. Topics and areas of transmission that I usually do not comment on now have been brought forward along with many additionally new topics that are also not mainstream. While these concepts from beyond were experienced, many of these ideas are not documented by others in the main stream spiritual community. Some Yes...and some No. Again remembering the message, I'm not the one writing this. There are messages or modalities touched upon in this book that I have approached from a Soul perspective bringing unique additions to the overall concepts and ideas.

If you've enjoyed reading or re-reading some of my other books, hopefully this book will be a true 'thought provoker' and bring some fresh answers and ideas for you from within your connection to the All in the Great Beyond in new and interesting ways. In this book we're delving into new concepts and ideas as we go for more of a quantum leap as well as exploring the internal energy and consciousness within the biodynamic physical body.

As we bring the messages through for you in the following pages I truly hope that this is a bridge that can help take you from where you are now on your journey to a new and exciting place. And maybe even that same moment of inspirational excitement that our ancestors experienced as they stood on a cliff looking into a fully lit night sky of stars and pondering (while internally knowing) 'This is the origin of our home. This is where we came from.'

As I channel both practically and esoterically the information in this book, some topics will naturally unfold for you, while others might challenge your old school of new thoughts and even some of the new age ideas concerning what we've come to believe and accept as our 'agreed upon' group or personal individualized truths. With a lit torch of inspiration and curiosity let's embark on this journey together.

Best Regards & Many Blessings of Light!

Bill Foss 3/03/2026

Transformational Information

Keepers of the Akashic Records

Messages From the
Keepers of The Akasha
the Ascended Masters
& the Arc Angels

By

Bill Foss

The Akashic Records Center
Institute For Higher Studies

Preface

QUEST FOR THE AKASHIC RECORDS

I'm sharing stories with you about my journey in hopes that it will give you insights and inspiration for your own growth. This is the story of a relatively normal guy about the inspirations and seeming obstacles along the life's path that found me immersed in continued spiritual studies and as a result of finding the Records, discovering my mission or Soul's Purpose. As I first found my way to the Records, I became inspired to continue the practice of accessing. I really wasn't thinking in the moment about the idea that I was soon to be encroaching on a life long study and pathway that would include writing, teaching and helping others in multiple different ways through insights given in my session work, presentations and workshops.

The one dynamic that kept me enthused and engaged with my studies and practice was the inspiration that I some how was already connected into the Akasha, as we all are, and all of the time, yet vastly unaware of it most of the time. But this was different. This was a past life connection and a heavenly in-between life connection. It was a strange and enigmatic elusive knowing that was somehow beyond my mind to consciously figure it all out and give a placement to it within the rest of my spiritual studies. As I continued my interactive journeys into the Records, I had a knowing and an unknowing (as if) that I was connected directly there in some major way beyond my understanding. In retrospect that deep connection that happens for me was unconsciously guiding me along in my beginning stages.

There wasn't anything calling to me at that point except my mundane life as an artist and musician, which was colorful in it's own right. Though as natural at it as I was, it made things even that much more seemingly mundane. I sought more mystical experiences as I continued journey meditation practices. It seemed like I had more time because it wasn't time for me to do anything major at that point, the writing, teaching and session work hadn't shown up yet, even as an idea. So while continuing to have meditation experiences, I was currently practicing spirituality while running an art gallery business, though when I found the Akashic Records I knew I was onto something big. That this was it. The thing that made the rest of my studies all seem to make sense and fit together.

So this distant knowing that I was somehow connected in a big way kept me diving, or rather jumping, into the Records and having more visions. The more successful meditations that I experienced, the more information I was being shown about my past lives. To some degree it was a bit overwhelming because, though as an artist and musician, I had entertained many delusions of grandeur (the ego). They were showing me some interestingly vivid and important spiritual lives across the past horizons of history where important things had been achieved.

I share this with you now here as one of the earliest steps that I went through and where it was going to take me. I was so attracted to the Akashic Records that somehow somewhere deep inside I had a notion that I was going to be using it in working with others, though I hadn't even consciously entertained it, even in the slightest. And yet I wasn't directly concerned with that, in so much, that guides and teachers kept hinting at the possibility for me of going into teaching. I never really took the bait on that because I humbly didn't really think I had anything to offer, though I had been studying metaphysics profusely for the past few years at that point.

It wasn't until an early teacher and friend asked me "So just what are the Akashic Records?" as we prepared to have dinner at her house while she was cooking. It wasn't until I gave her a 30 minute explanation did I even have a glimpse of knowing that I was to teach. In that moment there was a direct transmission of energy coming from me as I spoke about it. I could noticeably feel it. From that time my life started to change. I went out on the road travelling from place to place giving talks and workshops. Thankful for the good will of friends who would host me during my stay in their town, I made my way across the US for workshops, gatherings and expos.

As I continued to work at expos with individuals in readings a great field started to form around me. So much that it was noticeable to other practitioners and people who would come to work with me. I enjoyed helping others, using my gifts and supporting my way across the country.

The patrons always came to work with me continually because they could feel the energy of the Soul Field as I was connecting up into the Akashic. As I opened a Soul field of energy into the Records people would feel it and gather to have sessions. It also caught the attention of many of other psychics and healers and they often commented on it to me. I truly felt that I was there for the people who sought me out in order to offer healing and messages of virtue and life path guidance.

I was used to a social dynamic from the music industry and the fine art world for decades. Competitive personalities and attitudes were always around in my earlier careers. There was always a better musician or band and there was always a better artist or gallery. We as humans generally tend to be comparative in society as a way of recognizing our own place within it all. And so occasionally there would be talk among others questioning what I was all about. Quite simply, I was the new kid in town.

As I worked at the shows, one lady that I had done

an exchange reading for really gave me an important perspective. She had explained to me that I had awakened at a master level without any previous reference to it in this life. It was an early reference to a natural gift that seemed to find me in life and I had sought understanding to why I was so driven to be doing what I was doing. I understand that within the spiritual world, for the most part it is questionable, at best, for someone to refer to their self as a master. It seems that usually those who do, are not. And yet someone had shown up to describe what was happening to me. I had awakened at a relatively high level and it stirred the curiosity of those around me. I'd been called many things in my life, but a master wasn't one of them. I learned to look at it as more of a masterful approach to the intuitive arts and of using the Akashic Records for messages and healing of others and not that I was a master.

One lady that I was reading for at a Colorado show kept leaning over and looking directly behind me. I was used to folks having all sorts of reactions to the work, so finally I asked her what was going on. She said "I'm looking at the two really tall angels standing right behind you." This has happened multiple times in sessions and in speaking engagements for someone to notice the angels and to often say something. Occasionally certain Ascended Masters would also make an appearance. Among them were Sai Baba, Jesus, Mary, and Babaji to name a few.

As I had set out on a journey, that simple message did bring clarity that would help me to go another several thousand miles without questioning my path from within. My early teachers were great guides and memorable, sometimes challenging with me which spurred me ahead. I always kept a place of respect and gratitude for them in my heart and mind.

I had eventually come to the understanding and inner knowing within myself that if I were teaching others to do what I do, that it could be quite possible that someone would blossom and open in even more profound ways than what I have experienced. It takes the pressure off my ego of having to stay in any sort of alpha position. Though I have, do and will exhibit a foundationally strong teaching stance when working with groups and individuals in order to guide the total selves of those present who have, not only the Soul they want to connect with, but also the egos that will confront, challenge or doubt their own abilities and the messages of a teacher.

Sometimes especially the accomplished psychics who not always, though sometimes, could exhibit a fairly smug ego and would challenge or double-check every message that was being given. I would have to ask them to turn off their gifts and stop double checking everything in order to more functionally be able to understand, process and learn.

So as the story continues, I had been guided to Vienna Austria for one of my first workshops. I also felt that I was being taken to a place from some of my past lives that I really resonated with.

An old friend Jerry had connected me with a new

friend, Bridget in Vienna, Austria. And as I was scheduled to go over there for the international Wesak festival to show my art and to perform musically, it would be years before I came to find out that Vienna was an energetic and even magical place for me, in particular, to visit. Upon asking me for a description about the Akashic Records my new friend, Bridget, in Vienna promptly asked me to speak to her weekly meeting group at a restaurant with an interpreter. This was formally my first overseas workshop with an interpreter and with many workshops abroad to follow.

After that I continued to teach several workshops in Vienna yearly and have travelled back throughout the years for more workshops. While I was there on my first visit I was sketching pictures in a drawing pad in my hotel room. I seemed to have drawn a hooded figure with a semi-stern meditative look on his face who was doing a mudra or hand gesture associated with the heart chakra and chest area. It was clearly a guide or a past life reflection that I was picking up on. After later meeting this guide, or other version of self again in the dream state, I became more clear about that self and even used the mudra energy hand position in my later teaching material.

My next reflection of the Higher Self: A few years later back in the US, I was picking up on a version of myself that was much energetically larger and was occupying the Great Hall of Akashic Records. So I started to creatively and energetically pick up on 3 versions of self associated with the Akasha and the other side. An earthly self that was practicing spiritually and creating on the Earth plane (me), then an intermediary self who was like a guide or a 'go between', a mediator (much like the drawing). This intermediary self also had a very distinct Pleiadean feel and look about him. And then the higher heaven version of Self that was occupying the Hall of Records. A great being of light and very tall. I never really questioned 'how could this be?' but took it as part of creative and direct knowing that I was being shown.

I continued my day to day practices and it would be years later when I was shown that I could actually shift between the different selves and that I already was unconsciously doing this without knowing it as I did my spiritual and Akashic work. Another part of this very early on in my awakening stages as I studied was that I felt unusually alone, thinking and feeling that my guides were very far away from me. In relation to this set of guides as different versions of Self and my distant placement on the earth plane, I was right.

My next reflection of Higher Self: Once while visiting friends in California I had a dream vision while awakening close to the ocean. I was out of body in a peaceful space looking up when I saw a very tall being in a blue and silver metallic hooded robe float down from very far above. Very clear and up close with beautiful blues, purples and gold in his presence and garments or robe. There were almost tangible patterns of energy swirling around him from head to toe. As he floated straight down slowly I floated up to meet him. Astonished at the presence of this grand being I was telepathically asking him 'Who are you?' He was answering me back in a telepathic way beyond words that said "I'm your guide, I am you." He wanted to show his self to me and let me know he was there. Then he floated straight back up and disappeared as I woke up calmly very early in the morning. His presence was male oriented though the true nature of the being was androgynous. This was a direct confirmation of my higher self that I had channelled in the drawing years earlier.

On another occasion while travelling across the South western US. I woke up to find a grand glowing white hooded being very peaceful and silent standing over my RV as I awoke early in the morning. As I would be on my way to an expo across the country that day I would ask that higher version of myself to hold space and create a protective, interactive and intuitive vortex and presence in the booth space I would occupy days earlier.

Do all of us have multiple higher selves? Not all. Do all of us have higher selves? Yes. In my case there are different versions of stages of me from the Akashic Records as I step down to the Earth plane to incarnate, create and ultimately fulfil my Soul's Purpose which is to teach, heal and guide through service.

As I had first started learning and practicing to project my consciousness up into the Akasha to receive information, I realized the dynamic of this grand place being so far away. I also realized my continual place there in between lives when my earthly self would be absorbed back up into the Hall of Records. It was very far away. I always felt this. People would ask me, 'How can you read the Records in such a noisy expo hall?' I would answer 'I don't know, I just do'. Though I did know on some level that I already had a direct connection with the Keeper of the Great Hall and I would go straight up and leave to look into the person's Great Book for whom I was reading. So I would be in two places simultaneously. It really took a big distraction or a lot of noise to interrupt my connection.

So these are some of the details I have experienced in my past. I have learned to work with more energy protection and clearing over the years as a result of not initially understanding a need for it. This is something that I recommend to everyone now. You can find it in your own ways. Whether it's certain crystals and minerals, salt, prayers or a talisman. You can always reach out to an energy healer for healing and clearing sessions. It will be a part of your learning and understanding process.

May you find your way onto your True Path that lets in the Light of your Soul. I truly hope that this book will help you discover some deeper insights about your own healing, the Universe and your Soul's Journey in relation to the Akashic Records.

Original Sketch of my guide from the vision in Vienna, Austria

Chapter 1

Keepers of the Akashic Records

As we find out about the Akashic Records the words themselves seem to have a charge of energy. We may be studying other forms or elements of spirituality when we find out about the Akashic. This often peaks our curiosity. We have to know more, so we seek out information online or a book like the one you're holding in your hand. Some of us may find the term and just have to know more about what it is. Others may be curious but based on their own superstitious nature or a polarization around the dark and light of life they may have reservations about investigating or contacting the realm of the Souls.

The Akashic Records are timeless in the Eternal Now of the higher dimensions and by our standards of perception much more constant then we are used to here. You might say unchanging except for the constant entries being recorded into the Records. I have often found myself talking to people who wanted to know more about what it is. When I explain it to them they may sometimes say "Oh yes, I've been there, the great library." Either in dreams, in a vision or during an NDE (Near Death Experience). I often tell this group or individual that because you've been there you can have an even easier go at accessing into the Akashic realm and your personal Soul Records or Book of Life. It will be second nature for these individuals and I often encourage them to give it a try by connecting to the Akashic in dreams or in meditation.

You don't have to be part of this group of people to gain access. Recently I had a client who after practicing and getting consistent entry into the Akashic say that she felt guilty or remorse for not journeying up to the Akashic more often, as she had stopped her practice for awhile. This brought up an interesting message: It's okay to take breaks or even not to go there for a long while. The Akasha exists because we come down here to incarnate. The Earth reality is the counter part of the equation. As we come down and enter into a sentient body with senses and an ego. who is based around survival, we forget our own divinity for a moment in eternity as we start anew in each life to process, observe, react, respond, create and to love.

So the Earth plane is just as important in being here as it is for the perceived idea that we've got to get away from here. When we start to awaken, or at least to know that there is way more in reality than the Universe is letting on, we start to wonder if we're in the right place by being here. The answer is yes. The Earth plane is part of the Akashic and Soul's incarnation process. We are sent down here to play out a different variations or versions of our Soul's purpose. You might think of the Akashic Records as the inter-dimensional international processing center for incarnation of the Souls.

This individual spent most of her life jumping out of body during sleep and flying as far away as she could. It was her way of running from the traumas of the past. I'm sure there were many colorful experiences in her journeys. But it became tiring at some point because it was ceaseless and was not allowing her to rest at night, awakening feeling tired or blown out from sleeping. She had been aware of my work for a long time as we were friends from before I started teaching. She tried the Akashic journeys a few times but for the most part left it alone.

In one conversation, I expressed that because she was disassociating during sleep, and that maybe she should try using a *target location* of the Akashic Hall of Wisdom. This would be a safe and protected place for her to project out of body to because it is full of healing energy and it is a well protected fortress of light. When she started practicing she immediately got right in. Remaining present to the ethereal surroundings for the next few visits, I then invited her to find her Book of Life or request to start viewing other lifetimes which she did with colorful and clear success. I share this story here in the book for anyone who has had challenges with disassociating through their energy body. When you project to the target location of the Temple of Souls, you are coming to a place of peace, healing and love. And you start to connect with your Soul body instead of just the astral body which is the part of you that likes to venture out to cosmic travel the astral worlds, different dimensions or just the Universe in general.

The Keepers of the Records are Great Divine Beings angelic in nature. The Akashic Hall is a sacred place and they are high level beings who have been chosen by the high councils representing the Light of Creator God to oversee the transformation, the evolution and the incarnations of all Souls visiting planet Earth. There are so many diverse Souls coming here from other distant galaxies or civilizations within our own galaxy. When a Soul comes here to this planet to incarnate, their first stop is the Way Station of Souls, the Akashic Records.

As they meet with the Keepers of the Records they talk about what the world is like and what they would like to do. How would they like to create and love through experiences on the earth plane? What parts of the world would they like to live in or visit over a lifetime? Would they like to have a family or work alone? What would they like to work on in developing the character of their Soul? On the Earth plane we develop character as we are enjoined with a physical sentient body that's processing all of the many thoughts and emotions of Earth life through the mental and emotional energy field bodies that each of us have. We evolve more quickly in defining our character by incarnating here rather than just sitting in heaven, even though the vibrations there are much higher and more pleasant in nature.

The Keepers of the Records are well versed in developing overlapping Soul mate intersections on the Earth plane. Many of us have been here for so many lifetimes already that the aligning of multiple Soul mates with each other through birth and life path positioning is quite a great planned equation. Requiring multiple stages and overlays of projected lifetime reality paths to arrive Soul mates or karmic Soul traveller friends to positional moments of meeting up with each other.

It is easy for the Keepers to look at a group of Souls through the Records and to divinely calculate when, where and in what ways they will work with each other. Incarnating here on Earth in and of itself is the great miracle. Life itself here is the great creative loving miracle. How we choose to live that out is up to each individual Soul prior to incarnating, though also influenced by having an ego on the Earth plane. The balance of ego and Soul is the ultimate dance of incarnation. Too much in the ego tends to create karmic situations that, while it seems we are here to live and to experience, can generate far more than we expected.

Becoming aware of the Soul while we are in human form is the ultimate expression of influencing our lives, world and reality here. When the Soul is more connected and radiating into the physical body it can balance and even over ride the ego's sometimes monotonous and ongoing schemes, bringing a greater sense of balance and inner/outer connection with life itself. We can access the Soul through connecting with the Akashic Records and our Book of Life. In fact this is one of the only ways I have found of actually having a more tangible, knowing and sensual experience with the Soul itself. There are various other visualizations and meditations out there for connecting to one's Soul body, but then what can you do with it and where can you go from there? I recommend the Records as a way of connecting into your Soul Book, and the last place of origin prior to coming here because you've been there so many times already. Visiting the Akasha naturally activates your Soul body and starts giving you ways to work with it continually through the mind and the senses.

So who are the Keepers of the Akashic Records? Are they angels or Souls between lifetimes or masters? Yes, they can be any of these. There is a Great Keeper of the Records which you might think of as sort of a Merlin figure. Flowing robe, long white hair, jovial and loving, studious and mystical, and always working on something. There are others as well and they work together as a team. But they may show up for you in the way that your mind is ready for, or you may be there with your guides or even alone for the experience.

There are Individual Soul groups, the Soul grouping of humanity, the Soul groupings of races and the Earth civilization as a whole. This is a top priority position for directing the Souls through the Great Tapestry of Humanity. Much like working with one Soul in it's journey to the Earth plane to connect with many Soul mates and live out their life and Soul purpose, but more on group levels of humanity. While the Keepers can guide the individual Souls, they are also working with great groups of Souls simultaneously.

Do the Keepers of the Records ever come down to the earth plane to incarnate? Yes. You will find librarians working in the Records who have had their many lifetimes on the Earth plane and they are back in the Records to work being of service to everyone who visits here. Sometimes the Souls still having their incarnations can take a break and spend time in the Records to help others. Sometimes their lifetimes are finished but they may come down to do a special job of sorts in a lifetime that needs to be accomplished. I have met many of these other Keepers here in human form doing their work, some of them are even Akashic counselors helping others here through giving messages and guidance for their lives.

Sometimes I will see Ascended Masters coming to work with someone in the Great Hall or they may be given certain Souls to work with. They can even come and request to work with a new Soul as a student or to reconnect with a Soul that they have been connected to in other previous lives. The Angels can do the same thing. They may be assigned to certain Souls throughout all their incarnations or just for a lifetime.

I have even seen Pleiadian guides in the Records as they have a remarkable drive to be of service to this world by helping the Souls and also helping our worldly Earth plane evolve. Not all deities visit the Official

Hall of Records. Many have their own meeting places, ashrams or gatherings on the astral plane or another planes of existence. For many of these beings it is more right and fitting for them to have their own ashram or hall.

Keepers of the Records visiting the Earth plane can be very normal in appearance. You may recognize them right off by that distinct librarian persona or something energetically different about them. Many of these people are either aware of the Akashic or will be drawn to work with it in their lifetime. I've come to realize that I'm an incarnation of one of these Keepers of the Records. The celestial Soul that is above split off and came down once again into my current bloodline. Seems a bit lofty, don't you think? You might think that it has certain perks, but let me assure you, I feel emotions and have thoughts just like everyone else here. When I'm connected in though it's uncanny and there is Soul healing energy that is flowing from the Great Temple of Souls into whom ever I'm working with. That part is pretty awesome.

I've dedicated my life to this work and while this is book number 10 there are still more to come. As a result of my work in the Records I became aware of the many lives I've lived with several of the Ascended Masters while they were in human form. I also have a working model and memory of the Angelic Realm as I spent time there as well. There are others on the Earth plane and I can always recognize those who are earthbound versions of Angels. You could be one! An Arc Angel may have multiple working incarnating parts of their selves here on the Earth plane at any given time. Soul energy works much differently than we previously imagined. It became very apparent to me very quickly. I have met people down here who are direct aspects of Michael, Metatron and Ariel to name a few, as well as other versions of Gabriel. It takes one to know one. How is this possible? It just is. Angels manifest their presence invisibly in the fields of the Earth, or even at times a physical appearance. They also often use our energy fields or come through us to help others.

So how can it be that I have the memories of the Keepers of the Akashic Records and the Angels and the Ascended Masters? I remember many of the incarnations and lifetimes with them through the Akashic Records. I also find them coming through me when I'm working with others or out in the world. This dynamic happens from opening the gateway and calling on them while I'm teaching, working with others or doing healing work. Everyone of them was a normal guy or gal just like you and me. Their presence is testimony to all the previous work they did. This is where we are headed.

So who are these great beings? How long have they occupied the Great Hall of Records? Where are they from? And what does it all mean in relation to the way that the Records work and are kept?

CHANNELLED TRANSMISSION FROM THE KEEPERS OF THE AKASHIC RECORDS

We have always been and we shall always be as Souls. Constant in time and space relative to perspective and understanding. Not all things have a beginning and an ending. Our world, our reality, our dimensions are more constant. We hold space here for all evolving Souls. And there are other versions of the Souls that inhabit the Records. Earth bound versions and Angelic versions. Other versions of these beings that occupy the hierarchical structures of heaven. They deal with all influences and aspects of Soul visitations within the Records. Soul energy works much differently than the Earth reference of linear perspective and separateness.

The monitoring of Soul evolution and the monitoring of Soul passage ways and their life paths. The staff positions in the Akashic Records are protected. The Light that is emanated and radiated here is powerful and is Divine. Pure Divine Light Source Energy. Holding space as a beacon for all the Soul's to come home. A beacon of Light unto itself is the Great Hall of Records. For eons we have worked and existed here. Before us there was no time. We were naturally created along with the heavens. We were created in order to have the capacity to house and to process great numbers of Souls in their karmic evolution coming out from the Great Collective Soul or God.

As any Keeper of the Records is called away to do other work of the Elohim, another Great Keeper, another Soul of equal Light, Power and Divine Intelligence is assigned. In the same way that a walk-in or walk-through on the Earth plane would work, but much more seamless as they step through into the position, into the Being, into the robes of light.

The white light is being anchored by Creator in this Divine place. Our job is to not only process the evolving Souls and the karmas of revisiting reincarnating Souls but also to teach the higher evolving Souls. Teaching and dissemination of information through light and reverence, they all come here. The Ascended Masters, the Arc Angels and the Souls that are transitioning into the Angelic Realm for learning, healing and teaching to receive wisdom and understanding. Wisdom messages and teaching are given orally but they are also transmitted energetically. Peace, Love, Creativity, Abundance and Harmony be with you on your journey. We will see you in the Records.

Lyra
Cygnus
Pleiades
SACRED
SANCTUARY
OF TRUTH
COSMIC LIGHT
LIGHT STAIRWAY
Conceptual
Paths of
Fate
Pleiades
Lyra
Cygnus
Thread of
Destiny
Pyramid of Giza ·
Akasha Link
Mayan K ul ·
Star Calendar Records
Parthenon ·
Wisdom Anchor
SACRED EARTH LOCA-
TIONS OF THE
AKASHIC COVENANTS
Stupa · Mind
Dharma Archive
LOWER ASTRAL ARCHIVE
A DENSER VIBRATION

Chapter 2

DIFFERENT LEVELS AND LOCATIONS OF THE AKASHIC RECORDS

There are so many ways that we look at our reality and practice at perceiving the Earth and our home solar system within the galaxy while observing the cosmos as she floats through the Milky Way. All the while our galaxy revolving around the Great Central Sun every 26,000 of our years (the Grand Procession). As you are awakening to the Akashic Records and it's levels you are also awakening to so much more. This is truly a great time to be alive.

As we came from the stars to inhabit this world we continue to arrive en masse at the Great Way Station of Souls, the Akashic Hall of Records, God's Library. As the cosmos was formed and God expanded out creating the stars and heavens, the planets and celestial bodies, The originating Soul or breath of God also expanded out in all directions as individuated Souls. Pieces of God sent out from the Original Moment of Expansion to look for home worlds, such as ours, and other inhabitable places among the stars. God experiencing His/Her Godself through the energy light orbs of individual Souls looking for inhabitable worlds to perceive, observe, experience and to help to evolve all life in the Universe as we know it. Evolution is ever changing, continual and persistent.

As we continue to study the Milky Way Galaxy scientists using the Kepler telescope system have theorized that a staggering 300 million planets may exist in our galaxy alone! With approximately 3,000 inhabited planets confirmed at the time of their published studies in 2022, and with an expected several thousand more inhabited planets to be confirmed soon after.

Scientists have also approximated that there are 10,000 billion billion observable planets in our known Universe.

When we are working with the laws and properties of manifestation and the mirror of the mind, the Universe responds like a great hall of mirrors. Could it be that astronomers are looking at parallel versions of realities within the holographic Universe as well as the expansive local space-time reality Universe that we are physically and energetically connected into? Neighboring Universes, if you will.

As the Universe was created, all of the living inhabitable worlds have an Akashic Records system developed as part of the working mechanism of that living world. The Akashic systems are etheric inter-dimensional and atmospheric bodies existing as a living working system to support life and the transmigration of God's individuated Souls coming to and from each planet. Learning, evolving, experiencing, interacting with other incarnated Souls, and the planet they're inhabiting, as the whole interdependent dynamic continues to flow through space/time on each inhabited planet within the collective of galaxies, solar systems and planets. This is an overview of what's been developed not only here on our world, but other worlds across the cosmos as well.

Looking into the inner workings of Mother Earth's Akashic Record system is more than enough to keep our mind, heart, body & Soul, of this one incarnation, busy learning through experiences for a lifetime.

As part of the etheric body of the Earth, the Beings of Light (us), are constantly drawn to the Light of the higher dimensional light body of Mother Earth. Harmonically we find our way into the heavenly realm through the Akasha in the outer atmospheres of Earth. Whether we are coming from the Earth in transition out of human form back to our pure Soul body or coming back down into another life. We could be waiting to go elsewhere in the cosmos. Or we may be coming from the great beyond as a newly incarnating visitor.

It is here in the Akasha that we begin to know our surroundings as a being of Light (Soul) in a soon to be enjoined form and in the greatness of the Light of the many other Souls incarnate together on the Earth plane.

As the Keepers guide us into a suitable earthly life, helping to find a harmonic match with a mother and father which will help us to fulfill our Soul's Purpose on the Earth plane by projecting us here, then birthing us into 3 dimensional holographic reality.

The interactive Living Library of Souls system is quite intricately complex in it's connections, levels, and locations. It is quite enough for us to stay focused primarily on our own Book of Life directive and to continue our karmic (cause and effect) driven Soul evolution development program. While it is also beneficial to give awareness to the great Souls and Great Teachers beyond the veils that have been working in the Records since the Earth's beginnings.

There is a quantum aspect and a holographic aspect to the Akashic Records system. There are many different levels and locations through different inter-dimensional doorways that allow Souls experiencing at different levels and intervals to uniquely participate as they are guided into the right areas of the Akasha. It is then for them to visit with the Keepers and Angelic Guides and then experience through incarnating lifetimes on this world or different worlds depending on whether they are coming (staying to incarnate) or going (leaving for another world).

The first notable reference to other Akashic levels becomes realized when individuals or groups find out about the Akashic Records in detail for their self study and visionary practices. While some may have the idea that they will be able to get in or 'hack' the Records for whatever level of psychic information they are in search of, mystery ensues. There are always those who are not interested in doing the work but want the benefits. Hacking into the Akashic for ego based pursuits of quandary is not harmonically allowed. These inquiries land an individual in one of two places or categories.

There is another simpler replica of the Great Library in the 5th dimension or astral realm for spiritual, psychic or Akashic dabblers. Here they can look into and loop through their own vibration and levels of inquiry while never really being able to get to the higher levels of information. They will be shown simpler versions of information and possibly even dream like answers that their ego is looking for. This happens when Souls may be compromised by dark beings looking for a way in, or under dark extraterrestrial influence. Or maybe they just aren't ready to step up into the true spiritual work yet, so they are shown the beginner level.

You might say that this is a self protect mechanism of the Great Records system. In this way the highest level of knowledge and wisdom cannot be misused. For all practical purposes such egoic inquiries would not be able to translate, understand or make use of the higher knowledge and wisdom anyway. The other category is that the individual may get in and simply may not be able to get the information they are looking for. It largely depends on the Soul, the level of that Soul, and the types of information they are requesting.

The Akashic Records exists first and foremost for the purpose of the growth and evolution of the Souls' progressive journey as a fingerprint of God visiting the Akasha itself and the Earth. When people get cornered into their sometimes ego and drama driven ideas of why it's okay to use the Akasha in conjunction with certain other rituals or for militant purposes (as an example) They will be redirected to a lesser version in another lower or parallel dimensional location. This is sometimes referred to as the Hall of Unofficial Records. There, they will meet many bureaucrats both visiting and working amongst each other, looking for answers, versions of the truth, or even competing with one another in some way to get something maybe even a promotion. Think of it as more of an astral courthouse or a city hall setting.

Maybe an individual who is rooted more in the lower aspects of ego who is competitive or confrontive would seek access to the Akashic and by virtue land there.

You have to be using the true Akashic Records for the right reasons to gain access which are self growth, evolution, a quest for developing wisdom and service to others. In truth this is the synchronicity of why we are all here having this sentient experience on Earth in the first place. We are here to experience love, creativity, each other and the Earth. While being individual, in appearance and perspective, on a deeper level we are all connected through the Soul groups' matrix and a great inter-dimensional quantum wave of the Universe. A net like rainbow colored frequency field layered with never ending, ever continuing holograms. Known in the quantum scientific community as the variants space while known to the Buddhist lamas and monks simply as the rainbow field. The higher monks and lamas use the rainbow field to transcend the Earth plane and activate their 'rainbow body' to become transparent and disappear as they merge into this field. The field is made up of pure Creative Essence and pure Divine Love which quite literally are 'one in the same.'

Negative ego based inquiries into the Great Hall of Wisdom, the Akashic Records, are dealt with in different ways depending on the person, moment and situation. Some may not be able to simply get in. If someone is trying to access for a lower vibrational reason, they may receive skewed, untrue, or colorized information. This is when the ego convinces itself that it is getting the messages that it wants to hear. Other attempts may prove unfruitful by not producing anything at all. Though sometimes they may access the Hall when they look into the books and the pages are blank.

Note: This can also happen to students with virtuous inquiries as well when they are learning. The same dynamic - two different reasons. Often times what someone is really looking for is more of a straight psychic based answer to their questions. When entering the Records you are looking at the Causes behind the Effects, which will answer your questions both literally and figuratively or symbolically.

When you are truly ready to ask the question within yourself "Who Am I?" the Akasha will give you access in new and interesting ways. As you continue to look into it with reverence and steady practice it will reveal more of who you are through visions, dreams and meditation experiences.

There are two primary levels of the Records that you will be entering. By entering in through the Great Hall as your gateway experience into the Akasha, you have access to the base level of the library which is interactively alive and expands at will once connected to it. Here you can ask questions about world history or people in our Earth history records. This level is expansive and has all knowledge of the world as a living library. It can be somewhat overwhelming to tap into yet interesting as well as revealing.

This is why we recommend being guided into your personal Book of Life to connect with details of your own Soul's journey. This will usually include other incarnations ot lives lived on Earth. Your personal Records will have All of your lessons, experiences, triumphs, failures, creations, Soul mates and the list goes on. Everything that you could ever want to know about You. Better to start your journey from your own Book of Lifetimes perspective reference within it all. Your personal Records will naturally have access to other times and places within history. Once you have mastered connecting into your own Book of Lifetimes then you will feel more comfortably equipped to look into the expansiveness of the rest of the Living Library.

The rest of the Library contains interactive rooms that are expansive on different cultures and subjects. You will feel more adept to using them once you 'know' your own self better. This interactive and expansively dynamic catalogue system is infinitely detailed as is the whole of the central Akashic Records Library.

There is a higher level that you can go into. An 'upper chamber' that you may find through advanced access when you are ready. This upper chamber is going to have expansive gateways to information concerning connections to other worlds, other star civilizations, other places in time such as parallel and alternate realities. The upper chamber is reserved for accomplished adepts in the study of accessing. Because the whole living library system is interactive, it literally makes no sense to try jumping forward into these unknown realms before you are ready. You can move forward into the next time period or phase systematically of unexplored territory with more grace and ease if you're able to keep a linear perspective as needed in accompany to the quantum streaming information format that can also be given. Every time that you move a step forward it will be mind expanding as it steps towards you.

So there is a certain amount of etheric grounding that has to be done in the whole process. This is natural. We could liken this to the example of having a premature kundalini awakening experience. It can be very ungrounded or uncomfortable if you are not prepared for it. Much in the same fashion moving into uncharted uniquely similar dimensional categories within the Records could have a similar uncomfortable effect. Better to work smart and be cautious rather than to throw yourself into the cosmic blender.

When you are looking into the Akashic Records through your guided journeys and practiced viewing, It is always best to go directly to your own Book of Life. If you are having difficulty in accessing your Book then you may want to practice connecting with your Guides and/or the Keepers of the Records there in the Great Hall. At times it is possible to have a lucid experience of getting there and then upon trying to look into their own Book, participants have expressed not being able to get into their Book of Life itself, or if they do sometimes the pages are blank, rendering them in the moment unable to read the entries and messages.

If this happens, not to worry. You are not being punished or quarantined by heaven. It just means that there is something within your energy bodies, mind or nervous system that may be pensive about what you might find there. Quite simply, this is the Ego. This is often the case if you're running into a subtle block. Ask yourself 'Have I shifted into my ego's desires of wanting to see things in a certain way?' or 'Am I afraid of what I might see or find there?' or 'Do I have my mind already made up about my past lives based on what someone else told me?'

The situation of not being able to gain access is, in itself, part of the study practice. This shows you different subtleties about the nature of the work. When you've gotten into the Akashic Library this is the first step. You may want to remain calm and rephrase your inquiries in a different way. Look at the way you're asking or the energy behind it. Do you feel a sense of urgency? If so, this is always the Ego concerned with it's own survival. I recommend that you return to your pre-journey meditation exercises and remain calm as to not stir any of the same or other subtle energy conflicts within yourself. Do your energy relaxation work and retry a journey or wait a little while and try again. But don't give up!

If you tend to, or do intend to look around in the Akashic Hall seeking to look into other levels of knowledge such as anything outside of your Book of Life here are some helpful guide lines. When you are looking at the world (Earth) and it's history through time, it's cultures, events in history or other similar subjects, you are not only looking into past physical reality. You may be looking at the world Soul or the group Soul of humanity. You may be getting information that is Soul oriented as opposed to only physical historical data. So things can be seen from different perspectives rather than what you think you were originally looking to see or find information about. You can see both perspectives if you ask more specifically.

Time is not what we think it is and reality is not what we think it is. If you approach your viewing from this standpoint you can get some really interesting visions and snapshots of information. Don't be surprised if they ask you what you intend to do with the information. They may already know that and are giving you the opportunity to answer from a higher perspective. Accessing the Records is primarily for the goal of Self Study and Self Knowing.

This includes your Soul's Purpose, which is related to your creative path on the Earth plane and your karmas or lessons that you are working through on the Earth plane. So if you are looking at the world's history records in relation to your practice of reading the Records and improving your Self Knowing you are more apt to be in harmony with receiving more information.

For example, If someone, as part of a local conspiracy study group, might be seeking out deep state agendas or dark alien plots, etc. the drama of the ego is being activated. Trying to look at this kind of information in the Akashic is useless on a Soul level. Either the info can be skewed or the person may be redirected to a local astral plane based version of the Hall of Records, more commonly known in higher circles as the Hall of Unofficial Records. There they would also find a fair number of bureaucrats seeking, using and managing information on lower levels. If someone is seeking information about their Soul connections to their alien home worlds, as another example, this is allowed in the Great Hall of Records because it is connected to their Soul's journey, and this is in relation to knowing of the Self.

Another important perspective is that of looking into or accessing the Akashic Records from your Soul body connection you will have greater and clearer results. Your intuition comes from your Soul field and as you look from your mind's eye in through your Soul's connection you will have little or no energetic noise interfering with your Akashic Vision Journey. You are directly looking through the Soul body up into it's connection to your Book of Life concerning information of a heavenly nature about yourself as well as the connections it experienced in different time periods as lifetimes in human form on this planet (Earth). If you are looking to access the Records directly through your 3rd eye, brain and mind, you may get energetic challenges from the nervous system in your body which is part of your physical self. Or you may be just fine in that regard.

This is because your physical self may have a fair amount of conflicts from everyday normal Earth life, people, places and things that can create conflicts when trying to view in. You may find that you are prone to want to ask more ego based questions or be in the ego's energy presence while you are trying to project your higher mind into the Records. Sometimes it is possible to think that you're accessing through your Soul's intuition and your ego could be running the whole show. That's why my Akashic Records study programs contain a lot of information about the Soul, mind, ego, nervous system and Self Study around Self Help topics and their levels. Because if you start to heal your life from the Earth reality here and you are accessing the Records you will get better levels of information and you are healing and knowing yourself from both sides; Here (Earth) and There (the Akashic Records). If you are trying to access the Akashic and not doing any spiritual practices around self discovery, energy clearing, expansion or bettering your existence on the Earth plane your experiences may be varied. You can still do it. It would be more like going to the Akashic Records in your dense physical body. Ask this question "Am I ready to meet God or the Angels and Masters in my current moment where I am now?" Is it your ego, mind or heart who's answering? Maybe it's time to do some work. Ask from your heart and the door will be opened. The main prerequisite is having no internal conflict or judgement for the best results.

If you project into the Records from your higher mind through your Soul body and bypass the lower mind and the ego, this is better because your Soul is already part of God. Your Soul already has the carrier frequency of the Akashic Records. And your Soul knows your heart. So you are one to several steps ahead because you are in alignment with the parts of yourself that are naturally accessing the upload/download stream from and into your Book of Life in the Akasha continually.

MESSAGE FROM THE KEEPERS OF THE AKASHIC RECORDS

Your Soul is a finger print of God. Your physical body temple is a fingertip of God. Coming out across the Universe form the First Moment, your mission is twofold. To explore strange new worlds. To seek out new life and new civilizations. To boldly go with other Souls to known and unknown worlds. And at the same time to remember who you are all the way back to your origins as part of the Divine Creator.

Moving forward through space time evolution, observing the Self in relation to the Great Self, and looking back into your history as an originating Soul and as a Soul enjoined in human lifetimes of the past. These perspectives bring oneness to the Great I Am of which You and your Soul are part of. This is not the only path to Self Realization and Enlightenment but it is an underlying template of Truth that will lead you home into deeper levels of awareness of Self. Through this evolutionary process you will once again find your way back home into the higher realms as you progressively finish the karmic cycles of reincarnation. You will finish the looping process and return to a higher order where you will then perceive new quests for your Soul in new worlds and beyond.

Your mission should you choose to accept it:

To experience Love and Creativity through Oneness and Individuality from the Cosmos and to channel that through your life on the Earth plane.

Lyra
Pleiades
Cygnus
DIVINE
PRESEMIC
LIFE
Ayren
Lachesis
(Mexsune)
Clotho
(Spin)
Atropos
(Cu)
YOUR SOUL
TAPESTRY,
WOVEN
THE
WORLD
TREE
OF LIFE
THE FATES
COSMIC
LATTICE
OM
ESSENCE

Chapter 3

YOUR SOUL TAPESTRY, THE TREE OF LIFE & THE FATES

The Great Soul Tapestry in the Hall of Records is quite astonishing to behold. Much like the rest of Heaven, the perspective of it's perception is multi-dimensional. People have reported during near death experiences that distances seemed to have a perplexity of both near and far depending on what you are focusing on. In previous books I have mentioned this same type of dynamic in the Akashic Records being both near and far with a semi-transparent quality depending on what you are looking at or focusing on.

A colleague once pointed out an Earth plane example of this dynamic: That of the Indian master Sathya Sai Baba who would come out to give Darshan (Divine Blessings) to hundreds of followers. No matter how far away from him you might have been he always seemed like he was connecting directly with you and it seemed as if he was looking at you personally. I believe this is an energy quality of the other side/heaven. The Great Soul Tapestry located in the Central Hall of Akashic Records also has this quality. A woman who'd experienced the afterlife in a NDE (near death experience) described it as large groups of people gathered waiting to walk in to Great Hall of Records. Even though a Soul was on the other side of a great crowd, once she connected with that Soul or person, (we still have our astral bodies at that point) it seemed like they were immediately standing next to each other. So this near/far distance anomaly seems to be quite different or variable compared to 3D reality where spatial perspective is more fixed.

When you are looking into The Great Soul Tapestry it tends to make you feel as if you are both near to it and far from it because of it's size. You are looking into a tapestry which is an energy diagram of every Soul who is incarnate in human form. All of the Collective Souls of humanity are shimmering with their Soul essence in this Great Tapestry which appears to be several stories tall and seems to disappear out through one wing of the Hall into Eternity. Though other representations of it can be seen (I have personally seen 2 others) It is possible to perceive multiple perspectives of your Soul within the Soul Tapestry.

You can see where you are now at this point in eternity. You can see where you are in relation to other Soul mates and Soul group members. You can see your progression in many ways as it relates to the whole tapestry of Souls, to your ancestral lineage, previous lineages and your immediate and extended family tree. At more in depth perceptions you can also see how you are connected to other Souls as family, friends and individuals that you have had, are having or will be having an influential encounter with. For those of us coming from the Earth plane, as we step off or step up into the Records and come to stand in front of this Great Soul Tapestry, it can have the ability to open or expand your mind just by looking into it. This brings the near/far qualities of perspective and perception online in a personal perceptive experience. This way of perceiving is quite normal or constant for the higher dimensions, Heaven, Divine Beings and the Keepers of the Records.

At every given moment they are aligning different Souls for future incarnations with other karmic or influential Soul players. With Souls coming into the game and Souls leaving the game they are working to align and connect each and every Soul with every other possible Soul interaction so that each of us gets the most we can out of each lifetime movie lived. That's quite an amazing dynamic. They are utilizing both quantum and holographic qualities of the Great Tapestry to do this.

And it is all done through Love. In this Divinely Intelligent energy you can feel the connection of the Great Oneness as you stand before the Tapestry. Some of us are not ready to perceive this just yet. It would be too overwhelming. If this occurs, not to worry, we just walk over directly to our Book of Life and start there. Once we have become accustomed to working in the Great Hall with our Book of Lifetimes, those who are ready are then shown the Great Tapestry of Souls. Another example would be the opportunity to look at planet Earth from Space. It would be quite grand and give us a unique sense of perspective as to it's immensity, sort of like being at the Grand Canyon or looking down at a city below from an airliner, though on a much grander scale.

When we are looking at the Soul Tapestry with the guidance of the Keepers we are, at that point, doing Soul work and reviewing lesson plans or even preparing for our next level journeys into this world or another depending on the quality of our responsive lives lived.

Another important element of the other side is The Tree of Life. There is a great Tree of Life for all of humanity, and each individual family bloodline has it's own Tree of Life. The Tree of Life is like the Soul Tapestry but for an individual family lineage. When you go into the Akashic Records you may at times see a Great Tree that seems to be exuding energy. Many people have also witnessed this in shamanic plant medicine journeys on the Earth plane while going into the Earth Records. Though, sometimes it is referred to as the Tree of All Life.

This is your Family Tree or your Tree of Life. The Tree of Life has all of the Souls in your bloodline connected to it. It may show itself to you as a tree with many lights on it. These are the Souls in your family lineage. Some lights may be brighter than others. These are the more evolved Souls in your ancestry. Often your own Soul may appear as the brightest light on the Tree. This can be to give you a point of reference, or it may be that you are indeed carrying the brightest light of evolution in your family heritage on your Tree of Life. Your Light is available through the lineage to light up the rest of the Tree. As they gave you life by birthing you into the physical plane, you then stream light back to them through the DNA.

Often times those of us who awaken are carrying the torch for the rest of the bloodline. While we show up as the brightest light on the Tree of Life we may be the first one to awaken in our immediate family or even in the family bloodline. As we awaken we shine our Soul's brilliance out into this reality as a light for our own life path. Our light and it's radiance, our Soul radiance, shines both back in time and forward in time while continuing to radiate out in this Present Now Moment.

As this dynamic occurs, our Souls energy radiance also shines out to other family members on the Tree of Life through our DNA. All family members and ancestors are connected together through their bloodline or ancestral lineage. This is often seen by healers or intuitives as a long line of ancestors standing behind you.

Your radiance naturally reaches them and starts to heal and clear them. When there is an overwhelming amount of baggage in the ancestry, as we start to open to this brighter radiance, sometimes we may feel it tugging or pulling at us. This happens through the family ancestral DNA markers. This can be a challenge as you need to continue to clear and heal them for their well being as well as yours. You may seek out a practitioner who can truly help you in this process. Ancestral Healing sessions are available at my sites.

And at the same time you may also seek to break away and move beyond and away from the current lineage and this is quite possible. Though you may want to consider some common form of balance in working with them as Souls. Sometimes if your light is taken away it could have adverse effects on the family causing more challenging situations, karmas or illnesses, etc. So by doing the ancestral healing and clearing you are helping them to balance as you step back from the family dynamics if needed or preferred.

As we are carrying the torch for our family, relatives and ancestors remember that we must bear that torch for our own personal well being too. Sometimes you must choose whether to take care of your family's needs or your own personal well being. This is a question that often comes up in our lives as we move forward on our own journey through the awakening process. To take care of the family or to take care of yourself. Or to do both. People who are taking care of family physically and/or living with them, whether being provided for, or providing for them, usually have a sense of physical family duty or obligation. Maybe you have moved on from your family but the DNA still tugs at you from a physical distance, from the other side or both. Ancestral Healing can be good for both of these scenarios.

When you see the Great Tree of Life in your inner Akashic vision or in the dream state, this is a moment of great Soul reconnection for you. There is great energy from the Tree being given to you in such a vision. As your Soul is multi-dimensional and as you look into it, the radiant spiritual energy is being absorbed and recirculated through your mind, eyes , energy body and Your Akashic Dream Vision (inner energy vision). This is truly an amazing and unique dynamic to behold.

You may ask to look into your Tree of Life on your Journey into the Akashic Records or it may appear to you as a special energy blessing in the form of your Akashic Dream Vision. Blessed is your Tree of Life, your Tree of Souls in all ways.

When you go to your Tree of Life you may meet another family member there. There may be more than one family member. They may come one at a time, a few at a time or the whole family may come to join you. They can even be waiting for you as you arrive. There is no preferred way, it's just different in every person's Soul Vision. It will be a great gathering of Love and remembrance. In the Great Love of Heaven on the other side, there is no lower emotional vibrations. So any petty differences or family quarrels from the past will not be.

You will remember why you all came together in the first place as a family on Earth. But before you go that deep you will revel in the happiness of seeing everyone again. This can go on for as long as you feel called to participate. This is a special time together because some Souls in the family will be staying in Heaven longer or maybe for good. Some family Souls may be called away to ready themselves for another Soul's journey on this planet or perhaps going to another star system.

So the time you spend together is invaluable and priceless. As you come together with them you may stay together as long as you like. The family may create a setting for a picnic with chairs for relaxing or tables. Even a game to play, or sports balls to play catch with. It is a time for love and gathering. In some cases you may live together for an extended period of time in a setting similar to Earth, though in Heaven. This helps to heal time lost on the Earth plane such as in a case where someone transitions earlier in life and never got to see their family, parents or grand children grow up through life.

Remember you are in the Eternal Now and there is all the time in the world (or Heaven I should say). After that movie is done the Soul or Souls will return to either living their life there in Heaven or may go their separate ways to start new lives. There are other times in Heaven on the other side and in the Akashic Records when you may visit with anyone from your ancestral lineage or family, friends and mates from other lifetimes even if they are not from a single ancestral lineage.

Imagine having a meeting with multiple families you've been part of and seeing how some of the same Souls are overlapping between those families and lifetimes as well. Imagine having a meeting with all of your lovers, wives or husbands from multiple lifetimes and seeing the overlays there too. Or your children, friends, neighbors and business associates. So visiting the Tree of Life can be a very dynamic, comforting and interesting reunion that brings love and healing for the whole family in so many ways.

Another element of Creation in the Higher Realms that plays intermediary to the Akashic Records is that of the Fates. The Fates are a unique 3 part composite based on the Divine Feminine aspect of the Universal. On behalf of your Soul and life they can oversee the creation of events and can also create events in your life movie as potentials that come in along the event horizon and into your personal Soul's Tapestry. They have access to a representation of your personal Soul's tapestry, which they work with, as well as other higher beings and yourself included, when you are ready. The energy potentials are then passed on to you metaphysically through your Higher Self from their higher plane and finally to the physical you on the Earth plane. You might look at the 3 Fates - Maiden, Mother and Crone as an ancient 'other version' or similar intermediary working model of the Records and is historically and culturally specific to some regions of the world. There was a time when feminine deities were the norm and even God and the sun were both thought to be particularly female.

Then God and the sun were switched to masculine deities in competitive success planning for a new world and generation. The true nature of God and our sun are both beyond male/female designation. The Fates can be found in ancient Greece, Rome, Egypt, and even as far north as Scandinavia and the Celtic regions. In our modern Akashic studies we look at the dynamics of You connecting through your Soul and up into the Akashic

Records and your Book of Life with an upload / download stream. While also working with guides, angels and sometimes masters to influence or inspire your path through life. This form of study was alive in Atlantis along side the Fates, and was also present in ancient Egypt, Sumer, Early Celtic Europe and Scandinavia. I visit the Fates not as often as the Akashic Records, but I do stop in once in a while to check on my Soul Tapestry weave coming down to them from the event horizon.

The Fates, while ancient in their place within our creation, are more like moderators of energy coming to and from a person and are also connected through their own system of antiquity with the Soul's Tapestry, the Tree of Life, your Soul and it's lifetimes. As moderators they are represented in different cultures across the globe. They all seemed to have had their own versions of the Fates. In medieval Celtic myth and legend, Ancient Rome, and the ancient tribes of Ester they were originally known as the triple goddess. This was the ancient way based on a feminine version of creation and deities. The first version of a spiritual triune God came from this model of maiden, mother and crone. Younger sister, middle sister, older sister or daughter, mother and grandmother.

It has been said in some metaphysical and astrological circles that the times in life when their presence in indisputable is around the times of your astrological Saturn return process. You can look up a detailed story of the Fates in relation to your Saturn return online. This had definitely been the case for me. One night after researching the Fates in depth and during my second Saturn return, in the dream state I found myself at a small retreat or workshop center in nature. A young blonde haired lady who was quite kind and agreeable (the younger Fate) was counting money (energy) out for me on a counter in the kitchen. She said to me " The others will be here shortly." (referring to the other 2 Fates).

As a while passed and I walked around a bit, remembering a few others outside the house, all of a sudden appearing in from of me was a tall pail complected dark haired woman with very striking eyes just looking quietly at me. This was the middle Fate, the Mother. She never spoke, her gaze was quite enough for a first meeting and energy recognition. Another note for you here, is that Masters, Angels and higher beings often play multicultural roles and positions on the other side. The younger Fate is also Gabriel's wife. The middle or mother Fate is also Mother Mary, and the older grandmother fate is also Freya, mother of the Norse gods.

The later model of triune deity representation that sprang from this through Christianity and Catholicism was Father, Son and Holy Spirit. This model is a historical graduation and is uniquely different and similar in ways though acts differently through people's usage of it for spiritual purposes. It poses a focus on divine masculine designation rather than the previous divine feminine designation, which would largely become forgotten over time, though still used today in some groups and parts of the world. As you think or feel into the feminine (triple goddess) and the masculine (triune God) you may find that both do have uniquely different qualities. Some qualities and functions are the same or similar and some very different.

The triune God has been perceived as an evolving spiritual platform that seemed to have different qualities for an evolving world. In a time of historical redevelopment, the new movement decided to reinvent and change everything. The watchfulness, guidance, protection and blessings of a Divine Father figure as Godhead, or the nurturing, helpful, loving, and protecting aspects of a Divine Mother or Goddess. Mother or Father, who should we choose? And if God was a woman for so long, then what would become of the men? Humanity was trying to sort that out.

They were still in service to the Great Feminine Power of historical deities and were perhaps seeking a new and different approach to the Divine. We as humans creatively always seek new expressions. We do know that this shift took place after the birth and death of Jesus which had a lot to do with the God designation, making an impact in many parts of the world. We will need to look deeper into the Records at world history and Soul dynamics of those times for a better perspective to lay our quest to rest between the Feminine or Masculine Gods of history and what was really going on with the Soul of humanity behind the scenes at that time. We can see that even long before this great change up took place, that in India it was common for them to recognize both Divine feminine and masculine aspects of God through their deities.

We can look at the Fates as great moderators of the Universe through individual and group lives and we can look at the Triune God in the same way. The masculine aspect of God could also be looked at as moderator, though usually looked at as a fatherly governing persona through history. The Fates were looked at as cosmic moderators for the tribes in a unique way representing the women in the family model. Raising and caring for the children, story telling, teaching and learning, farming, food harvest, food preparation, clothing, tools and then hospice care for the dying.

The Daughter is known for braiding the Soul thread. The Mother is known for interacting directly with the individual and their Soul thread as it comes through them from their Soul Tapestry. The Grandmother is known for cutting the Soul thread at the end of a person's life. Grandmother also has the duty of cutting and trimming away things that are finished in a person's life. There is nothing sacrilegious about the Fates they are just from an older spiritual system which predates Christianity. Like the Arc Angels they are beings of responsibility and power and have specific jobs concerning energy flow to and from peoples lives on the Earth plane as intermediaries. This is about you learning to work with the transference and management of energy.

In the next guided meditations we will work them directly.

THE TAPESTRY OF SOULS, YOUR TREE OF LIFE AND THE FATES

Guided Journey Meditation Preparations

In preparation for the exercises, I will give you some guidelines in order to get the most out of your practice and connection. In the first meditation we visit the Great Hall of Records and the Tapestry of Souls. In the second meditation we will visit your Family Tree of Life, your family and ancestors. In the third meditation we will be moving outside the Great Hall and taking a walk through nature to the Office of the Fates.

I will be giving you some new concepts here to work with on the other side. It will be in the higher dimensions and adjacent to the Akashic Records. The location and the dimensions will activate naturally for you as you connect in. In reference to working with the Fates on our Souls's journey into the physical world of reality, it is connected through multiple dimensions on it's way from the Akasha down into the physical world. This is one of those connections and can be a significant one for doing healing, clearing, protection and recovery work.

Some of the many questions you may have had about rewriting your own entries in your Book of Life can be answered here. While I do not recommend trying to do that because the best way to affect your Book's entries is by creating, healing and clearing for yourself and your karmas (cause and effect) from the Earth plane. Working with the Fates offers an intermediary step which will affect you both here in physical reality and in the Akashic Records.

As a reference, this work is divinely ordained and offers the extra steps in working with the Fates. As you come to understand the process, you may visit them specifically once in a while without having to go to the Records first, as it's all connected in the network of the higher heavens.

We have been taught that we have our personal chi or vital pranic essence within our bodies and energy fields and that our Soul body is an Infinite Being connected to the Akashic Records and the Universe. While this is all true, your Higher Self allots you a certain amount of energy each day while you are on the Earth plane. When you have things happening in life that need a workable method of passage, the Fates can help with that.

The advantage to working with the Fates is being able to work with agreements, karmas, curses, fates, vows, oaths and hexes, etc. in a multi-dimensional primordial way that may be affecting you from other people through time or other beings who would lay claim to working with you or against you. So if you've felt that you have trouble with any of these, you will find that working with the Fates can prove to be clearing because you will feel the burdens lifted from you. And as always, myself as well as other healers are always available to work with you, to give extra guidance clearing and healing in those instances or situations where you feel the need for intervention and assistance. The idea of external assistance comes all the way from the early tribes as an individual was designated to be the Shaman for the whole tribe. While it is true that we can all heal ourselves, some things are best attended to by another person as the healer.

If working with the Fates is new to you, this can be another interconnected and practical way of working for protection, clearing, and the releasing of any potential negative influences that could reach you on the Earth plane from a multitude of variable past situations, people or energy beings.

In my work I have often commented, when you've tried everything else and nothing is working, then go to the Records. Working with the Fates is another intermediary level along the same principle. It can be very clearing and protective for your journey currently and moving forward in life. Enjoy working in realms of the Akashic Records!

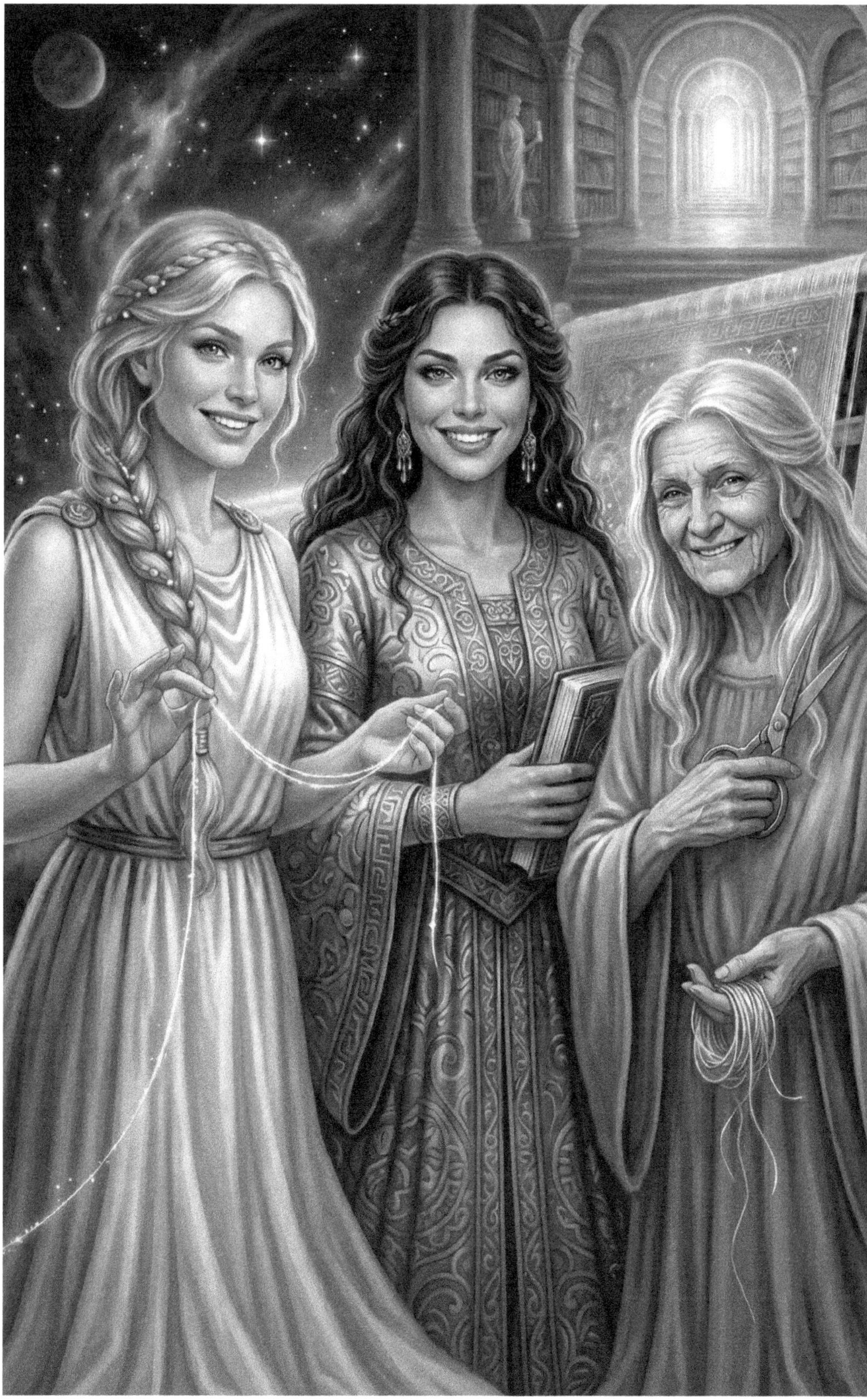

Meditation Exercise 1

Journey to the Tapestry of Souls

Guided Journey Meditation - Always remember to never do this work while driving a vehicle or using any type of machinery. You must always be stationary in a quiet meditation. Read the script and then close your eyes and visualize the journey.

As you sit or lay down getting comfortable, really let yourself start to relax more and more. Breathing in and out very long, slow and evenly. Just let all the cares and the concerns of the day fade away. Relax every part of your body from the top of you head to the tip of your toes. Become present with each part of your body as you relax that area letting it gently release.

As you continue to relax your heart rate slows down ever so gently... and now relax your mind with gentle peace. Just allow it to really slow down knowing that all is well and as it should be. Relax now your facial muscles, your forehead and your jaw. Allow your mouth, lips and tongue to relax. Allow your eyes, eyelids and your eye muscles to release and relax ever so gently. Relaxing your ears and your hearing. As you quiet and relax down even deeper taking the time for your body, mind and spirit, visualize a beautiful golden sphere of energy around your head and chest. A large sphere that extends up over your head. This is your Soul energy. As you see it, start to breathe this energy down into the top of your head.

Breathe the golden light of your Soul into your face. As you breathe the golden light in now coming down into your chest and heart. As you continue breathing the golden light of your Soul down around your body and through your body until it fills you up completely. Fully coming into your body lifting you up gently and supporting and soothing you in every way.

See now a beautiful portal of light coming from above as it comes into the room. As you continue to relax and focus on the portal of light, the more you look into it, you notice that it extends farther and farther up, higher and higher. As everything else in the room starts to become just slightly dimmer and fades back, the portal of light becomes more prominent and present around you.

As it continues to grow brighter it gently seems to be calling you into it's light. As your body starts to feel lighter and lighter now, you start to levitate and float up into this great Divine Light. Floating gently up, the higher you go, the more you seem to be able to view into this portal. The light extends all the way up through the heavens. As you see the stars and celestial bodies. You see a great field of white clouds just above you. They are glowing against the backdrop of space. As you move through the portal of light into the clouds, they are glowing with white light all around you. As you come out of the clouds you come to rest atop a sea of clouds and facing you is a very large Divine Temple Sanctuary. This temple is so peaceful and calm. It is emanating Divine energy.

There are 7 steps up to the 2 large wooden front doors which are trimmed in gold. As you walk up to the steps, you decide to enter the Great Hall so you take the first step up and then another.

1 - 2 - 3 - 4 - 5 - 6 - 7

You are now standing before the temple doors of the Great Hall. As you look up they are so tall. And there on the temple landing you notice there are two great lions, one on each side of the entrance. The female lioness is on the left and male lion is on the right. They are here sitting quietly guarding the Great Hall. As you reach your hand down to pet their fur, feel the softness of the texture. They purr and look up lovingly into your eyes as you greet each one of them with a gentle stroke of their fur.

As you turn now to the Great Temple doors in front of you, reaching out your hand as to open the door, the doors very slowly swing open revealing the inside of the Great Hall of Records. Just your intention was enough to open them. You see how amazing this place truly is. There is a gentle soft white Divine Light like a mist or a fog that streams in through the very large stain glass windows lining each side of the Hall.

There is a great arched ceiling that runs the length of the Hall. As you walk into the Great Hall and out onto the middle of the floor, the temple is so large that it's expansiveness allows your energy fields to increase and expand. You feel the greatness of your Soul and your True Self here in the comfort of the Akashic Soul Temple. You feel a peace that passes all understanding.

There are very large stained glass windows lining the Hall on each side as the cosmic star light of creation streams down in colored beams that shine patterns across the great floor. You notice now along each side of the hall there are book cases and tables with Great Books and Scrolls laid out. On these scrolls and in these books are written the stories of Creation. Of all that has been and all that will ever be.

As you look to the center at the far end of the Hall there stands before you the Great Tapestry of Souls. A tapestry which is 22 feet in height. It is shimmering with multi-colored Soul lights all woven together that make up it's entirety. To the left there is a shimmering holographic wave of the Tapestry moving out through the wall of the hall signifying all the Souls of the past who have gone on to other worlds and dimensions in the Great Cosmos. To the right is a shimmering holographic wave of the tapestry moving out through the wall of the hall signifying all of the Souls that are waiting to come in or that are on their way here. This shows us the greater past, present and future of the Soul's Tapestry activity.

As you stand there in the Akashic Hall of Wisdom before the Great Tapestry, there is now a great ring of golden light coming out of your left side midsection and streaming up in a circle 12 feet above you as it circles back down through your right side midsection. The energy of this great golden ring of light is flowing up, over and back around down through you in a clockwise direction. It is creating a great ring of high frequency perception for you as it activates your senses, your mind, your Heart of Hearts and your Soul field. It is also holding your ego in balance with your Soul field energy.

As you look now into the Great Tapestry notice all of the details of the Soul lights and their colors. See how they are stationary or seeming to be in motion. In the brilliance of the colors you may even see colors that you have never seen before. Notice now where your own Soul is in the Tapestry. As you watch it, notice the color your Soul represents to you. As you continue to watch, look now at all of the Soul lights around you. Some are very close and others are father away. We notice the Souls in the Tapestry that you are connected to, or that have an influence or impact in our Soul's Journey. Notice the colors of the other Souls around you in the Tapestry. Looking into the Great Tapestry of Souls can bring you a sense of being more interconnected to all other Souls. You are looking at the whole of Creation from the Soul's perspective. Breathe into this for a few moments.

Now before you the whole floor of the Hall turns into a great mirror. As the Souls around your position in the Tapestry light up, they all shine forth beams of light down onto the mirror floor. As this happens each beam turns into a golden sphere of light before you. Up from the mirror comes a holographic person into each one of the golden spheres. Each golden sphere is a Soul and each person in each Soul Sphere is a person that you know or are connected to. They can be friends, family members, present, past or future partners, Soul mates, or business associates. They may even be Souls that bring karmas to complete, heal and clear. Take a few moments to look at the people there before you in their Soul spheres. As you now move into your Heart of Hearts Love Frequency hold this love present within you as you look at the different people in their Soul Spheres.

Some of the people represent love relationships as you take a moment to heal past love relations, to fortify current love relationships or to open to coming love relationships. Look now at friends and family. What are the connections between you? What are you helping each other to experience? Take a moment now to look at business and creative people and their Souls. What are you helping each other to create, build or to finish? You may even see people that you will meet in the future. Perceive now what they are coming to share.

As you are here now before them all, from your Heart of Hearts project out Spiritual, Mental, Physical, Emotional and Financial well being in a great field of your Heart's Love Light. As it reaches them all in their own unique ways, there is Great Healing, Balance, Acknowledgement, Clearing and Abundance taking place on a Soul level that will translate through to them and to you on the Earth plane. Take a few moments to feel and experience this energy transmission.

As the process is now perfect, whole and complete, the figures become transparent and fade back into the mirror in the floor. Their Soul Sphere projections float back up and return into the Great Tapestry. As you now feel more comfort, clarity and full of love, knowing that all things needed for healing are well, whole and complete. And all possible and probable opportunities that are coming will be strengthened. Giving thanks here in the Great Hall of Records for the experience of working with the Great Tapestry of Souls, let's move back across the Hall towards a portal of Light.

As you turn around to walk across the Hall, you see several tapestries hanging over the front entrance of the Hall. They are moving with energy and pictures. These are some of your past lives. Take a few moments here to view into the images being shown to you.

As you now continue to walk across the Hall, you step into the portal of Light it transports you back down through the heavens and back down to Earth. Coming back into your home and your physical body take a few moments to reintegrate back to the here and now.

We thank the Keepers of the Akashic Records for facilitating this Great Soul Awakening within ourselves, our lives and our Earth bound reality. Om Peace Amen.

Meditation Exercise 2

JOURNEY TO THE TREE OF LIFE

Guided Journey Meditation - Always remember to never do this work while driving a vehicle or using any type of machinery. You must always be stationary in a quiet meditation. Read the script and then close your eyes and visualize the journey.

As you sit or lay down getting comfortable, really let yourself start to relax more and more. Breathing in and out very long, slow and evenly. Just let all the cares and the concerns of the day fade away. Relax every part of your body from the top of you head to the tip of your toes. Become present with each part of your body as you relax that area letting it gently release. As you continue to relax your heart rate slows down ever so gently... and now relax your mind with gentle peace. Just allow it to really slow down knowing that all is well and as it should be. Relax now your facial muscles, your forehead and your jaw. Allow your mouth, lips and tongue to relax. Allow your eyes, eyelids and your eye muscles to release and relax ever so gently. Relaxing your ears and your hearing. As you quiet and relax down even deeper taking the time for your body, mind and spirit, visualize a beautiful golden sphere of energy around your head and chest. A large sphere that extends up over your head. This is your Soul energy. As you see it, start to breathe this energy down into the top of your head. Breathe the golden light of your Soul into your face. As you breathe, the golden light in now coming down into your chest and heart. As you continue breathing the golden light of your Soul down around your body and through your body until it fills you up completely. Fully coming into your body lifting you up gently and supporting and soothing you in every way.

See now a beautiful portal of light coming from above as it comes into the room. As you continue to relax and focus on the portal of light, the more you look into it you notice that it extends farther and farther up, higher and higher. As everything else in the room starts to become just slightly dimmer and fades back, the portal of light becomes more prominent and present around you. As it continues to grow brighter it gently seems to be calling you into it's light. As your body starts to feel lighter and lighter now, you start to levitate and float up into this great Divine Light. Floating gently up the higher you go the more you seem to be able to view into this portal. The light extends all the way up through the heavens. As you see the stars and celestial bodies. You see a great field of white clouds just above you. They are glowing in the backdrop of space. As you move through the portal of light into the clouds, they are glowing with white light all around you. As you come out of the clouds you come to rest atop a sea of clouds and facing you is a very large Divine Temple Sanctuary. This temple is so peaceful and calm. It is emanating Divine energy.

As you look to the left of the Great Hall you see a beautiful outdoor setting. A beautiful blue sky with white puffy clouds. There are grassy hills and beautiful large luscious trees spotting the landscape. Nearby in the distance you see a lovely still lake. The setting is very relaxing and there is a gentle breeze and perfect temperature on this sunny day.

As you start to walk slowly across the grassy knoll moving around and away from the Great Hall, looking at the great outdoor nature scape, you notice that there is one tree that seems distinctly different. It seems to be silently calling you. As you walk over to this tree you notice that there seems to be a subtle glow around it making the tree stand out from the rest of the landscape and sky. The light of the sun is creating glistening little lights all over the tree. It is shimmering with energy. You have an intuitive feeling that you are supposed to connect with this tree.

As you come closer you notice just how alive the tree is. You can feel the leaves of this Great Tree breathing. It is so tall. As you walk over to touch the trunk of the tree, closing your eyes, you feel strength, grounding, love and healing gently flowing from the tree into your hands. You then start to see the faces of many of your family and distant relatives in your mind. Some that you may have never met.

As you open your eyes, you notice that here under the shade of the tree on the other side is a person standing with there back to you looking out across the beautiful scenery. As they are saying something to you, they slowly turn around and you realize who it is. It seems like it's been so long since you've seen them. The joy in your heart is accompanied by the tears of joy welling up in your eyes. This long lost family member is standing here with you now as you walk towards each other and have a hearty embrace! There is so much to talk about though telepathically you both already know. The conversation is good. Take a few moments to reconnect by having a conversation with them as you listen to their messages.

As you're both talking, over their shoulder in the distance you see another figure coming this way. As they get closer and closer you realize that it is someone else that you are very close with. As they're coming closer you both turn to greet them. It's been so long and yet it somehow feels that you've always been together. As laughter echoes and hugs of great joy are shared you both strike up a conversation with them. Listen to them and their messages as you stand together with them here under the Tree of Life.

As you've exchanged stories and laughter you now sense that people are approaching the tree from multiple directions. To your surprise the rest of the family is all coming together. What a wonderful day for a family gathering. As you meet and greet everyone there is now time to tell family stories and laugh. It is a time for healing and understanding through love for any situations with your family members that are in need of healing, forgiveness and love. Listen to the messages of wisdom, love, forgiveness and healing exchanges from your family. Take a few moments here to be present and receive messages from them.

You feel such a burden being lifted from the forgiveness and old stories of the Earth life together. There is a great sense of completion within you and comfort. Take a few moments to really feel this within your Heart and Soul. As the Great Tree welcomes everyone to it providing a comforting meeting for the family gathering. You realize that the tree is glowing again. Sparkling with light as the branches overhead come alive.

As the branches of the Tree of Life seem to gently stretch out in all directions, they grow longer as there now others coming from the distance in all directions to touch the branches of the Great Tree. As they come forth and gather around touching the branches as they say hello to everyone. This is your extended family. Great grandmothers and great grandfathers, great great grandmothers and great great grandfathers, aunts, uncles, cousins, nephews and nieces. As the ancestors keep coming all to form a great circle lining up around the Tree of Life.

This is a time of great gathering, understanding and healing for the whole extended family. Take a moment to connect with your distant relatives as you circle through the crowd greeting them with hand shakes, smiles and hugs. You may notice that some of them are glimmering or glowing with energy, showing you that you may have been that incarnated ancestor or that they have some influence with you. Take a few moments and listen to their messages for you. Notice which relatives may need healing as you send them extra love and healing.

As you now continue to talk on this endless sunny afternoon some of the family members are talking with each other. And some have gone off to take a walk by the lake. You may ask some of your family if they would like to walk through the courtyard behind the Great Temple connecting to the other Great Halls.

As you walk along the pathways between the great structures such as the Hall of History and the Hall of the Arts, you come around the other side of the Akashic Temple where you see the Fountain of Healing Waters. As you walk up to the Fountain you spend a few moments here with your family rejuvenating your Souls. All of the lifetimes have been a long series of Earth journeys and now the Fountain of Healing Waters breathes life into your Souls. As you feel the mist on your face the Healing Guides are also there to project Soul healing energy to you and through you and your family. Take a few moments to receive this healing.

Feeling refreshed and joyful in your Soul once again, you sense your connection to the Universe, all life, all creativity and all love. Thanking the guides for their healing and the healing waters you hold out your hands in the fountain and splash some of the healing water on your face. It is so refreshing and soothing as you feel calm soothing peace and clarity running down through your nervous system and body. As you turn now and walk back through the courtyard to the Great Tree of Life on the other side of the Hall where many of your family are still talking and sharing stories. As you take a moment to look out across Creation and take in the amazing energy of Heaven, you know you will want to come back and visit the Great Tree and your family when the time is right.

So know bidding everyone a loving farewell you know that you're always connected here through the Great Tree of Life. As you start your journey back to the front of the Akashic Temple where you will find the stairway portal to bring you back to earth. Stepping on to the stairway and easily descending through the clouds in the heavens you look out at all of the beautiful stars in the sky as you are coming back down to Earth.

As the Earth is coming closer and closer you now see your part of the world. And you now see your part of the country. Coming closer you see your state and your town as you gently descend back into your home and back into your body. You know that you have brought back with you the family healing and the energy of heaven. You have brought back with you the energy of the Healing Waters and the Great Tree of Life. Take a few moments to be present with yourself as you reintegrate back into physical reality. You may even decide to take a nap or head off to sleep which can further deepen the healing you just received. Many Blessings to You and your Family and your Great Tree of Life. Om Peace Amen.

THE THREE FATES:
INTERMEDIARIES OF AKASHIC DESTINY
GREEK MORIAI
NORSE NORNIR
CELTIC MODRON
EGYPTIAN UATCHEY
MAIDEN - THE WEAVER OF THREADS
LIFE EVENTS - THE WEAVER OF TALES
THE CUTTER - THE TRIMMER OF LIFE THREADS

Meditation Exercise 3

Journey to Meet the Fates

Guided Journey Meditation - Always remember to never do this work while driving a vehicle or using any type of machinery. You must always be stationary in a quiet meditation. Read the script and then close your eyes and visualize the journey.

In preparation for this exercise, I will give you some guidelines in order to get the most out of your practice and connection with the Fates. First, you will be relaxing down into a state of inner peace as we always do in preparation for the guided Akashic journeys. In fact you will want to be slightly naturally relaxed and stationary. I recommend doing this procedure in a sitting position. If you need to do it while lying down that's fine also.

Your Higher Self from the Akashic Records will be with you on this remote journey outside of the Records. As we go to meet the 3 aspects of Divine Feminine Power known as the Fates. We will meet them on the other side in a beautiful remote location. As you meet them, you may offer them a gift of some sort in respect. Next, our

Higher Self provides us a certain amount of energy per day which is like a currency. For the work you will do with them, we will give them an energy gift in trade for the work that we will then ask them to do. I usually give them gold coins representing energy. The amount is always the same and is a discreet acknowledgement only between you and them. You can ask your Higher Self for the amount of energy to be given. As we will then proceed with all questions and 'gifts' that you have 'paid' them for. This is a very old historic and traditional way of working. So now we are ready for this unique clearing and protection journey.

Understand this: You are doing maintenance clean up work with any people, beings and life situations which are not originally in your Book of Life in the Akashic that were set up for you to experience, but may come into anyone's life path as haphazard events caused by energetic interactions from any dimensions funnelling down to the Earth plane from between your originating point of origin in the Akashic Records, alternate realities and your life on the Earth plane. So this is your point of reference and working model for this journey. Enjoy!

As you sit in a comfortable upright position, either in a chair, couch, or cross-legged with a pillow tucked under your tail bone/buttocks to help you sit up straight while relaxing. Relax your hands on your thighs or knees with the palms up or down. Start to take some long slow deep breathes. Relaxing your whole body as you become still and peaceful. While continuing to relax down, down, down even deeper, feel your whole body letting go of any and all tension or stress. As you breathe in the light, peace and calm clarity....Breathing out the all of the tension and stress you've been holding on to. As your body is more fully relaxing down, down, down even deeper, continue to breathe in the light...and as you do, releasing all tension, fear or anxiety that your body may have been holding on to.

As you find yourself in a beautiful outdoor setting in nature walking along a trail in the woods, you are close to the base of an ancient mountain. The crisp smell of the very tall pine trees all around you is in the air as a gentle wind blows down the crest of the mountain.

As you stand here on the path, allow yourself to project your prayerful thoughts upwards inviting your Higher Self to come down at this time and accompany you on a special journey for a meeting with the Fates. As he or she gently and quietly floats down you start to move forward walking slowly on the path through nature. As you see the sun shining through the clouds down onto the path, you notice the passing patterns of shade on the forest and meadows from the clouds in the sky above.

Up ahead you see a gentle curve in the path and as you arrive at the turn you now look around the bend. Up ahead you see a small cabin nestled in the woods at the base of the great mountain perched there in nature. As you now walk slowly and calmly up to the cabin, You step up onto the old wooden porch and turn the front door knob. As you open the door, you see one open small room with a long counter the length of the room and a small window in the outer walls on each end of the counter.

Behind the counter you see three women looking back at you as if they have been waiting for your arrival. Your Higher Self is standing with you, behind you or over you. The youngest of the 3 is on your left. This is the Daughter Fate, a young lady in a traditional light colored dress with blonde hair. She is traditionally the most agreeable of the 3 and will usually give messages as prompts to indicate what is going on with all 3 of them. She is the one that creates the Life thread for your personal Soul Tapestry.

Next to her in the middle you see directly in front of you the Mother Fate. She is silent usually but with a solemn or striking gaze. A tall dark haired woman. She is the one who weaves the thread as it moves into your Soul Tapestry. The Soul Tapestry has a loom on the event horizon behind the house. She is the one who will be most direct with you, as they all can be, if needed, in their communications to you. To the right of her there stands a mature older woman with glasses and long white braided hair. This is the Grandmother Fate. Her job is cutting the chord at the end of your life. Not to worry we wont be doing that here. She does constantly trim away things that can be in your way or that are no longer needed.

As you greet them now, you say:

" Hello Maiden, Mother and Crone.
Younger Daughter, Middle Daughter, Older Daughter.
Daughter, Mother, and Grandmother"

Today I bring you a gift in honor of the great work you do with all Souls and the Universe. I also bring you money in the form of energy. (As we will now ask a series of questions for what we would like to purchase). Listen for any greetings or messages from them.

As you watch with your Inner Akashic Vision, take a few moments to see what the Fates put down on the counter in front of you. They will set everything on a mesa healing tapestry for you as you make each request.

I now ask to purchase any and all Karmas, Curses or Fates concerning Safe Travels. They may give you situations or objects as smaller representations of people, law enforcement, tickets or vehicles etc. as objects. They will place anything pertinent to your current life on the counter after your asking. Then you just move to the next question.

I now ask to purchase any and all Karmas, Curses or Fates concerning Legal Matters. Again take a moment to see what they are setting on the counter for you. It may be documents, people or situations, etc.

I now ask to purchase any and all Karmas, Curses or Fates concerning any Accidents of any type. Watch now to see if they place anything there before you. If they do place something there on the counter for you then you are not manifesting that, you are simply intervening with any would be challenging situations.

I now ask to purchase any and all Karmas, Curses or Fates concerning any people or beings which may have ill intentions towards me. Watch to see who or what they place before you.

I now ask to purchase any and all Karmas, Curses or Fates concerning any Curses disguised as Blessings. Watch carefully now as it may be a situation or object representing this. If there is nothing then no need for concern and continue. If there is something, you are now aware of it and your Higher Self will take care of it.

I now ask to purchase any and all Karmas, Curses or Fates concerning any Blocks to my Success in Life. Watching now again to see what they give to you.

I now ask to purchase any and all Karmas, Curses or Fates concerning any Blocks to my Relationships. Again they may bring you people, situations or symbolic objects.

Take a few moments now to ask about a particular situation that you may need help with.

I now ask to purchase any and all Karmas, Curses or Fates concerning ___________________.

I now ask to purchase any and all Karmas, Curses or Fates concerning ___________________.

Giving thanks to them for any and all objects they have placed before you on the mesa cloth. You may now pick up this mesa cloth with everything in it. As you do this in your mind, lift it up over your head and ask your Higher Self to transform everything into Pure Positive Light.

As it is now converted back into Pure Positive Light while you are holding it overhead, pour the Pure Positive Light back down into your self, drinking in the Pure Positive Energy. If there is anything that you do not wish to consume ask your Higher Self to filter it out and send it up into the Light of Heaven.

Giving thanks now to the 3 Fates, you may ask them for a message and a blessing for your Life Path. As you prepare to leave their little house in the woods now and starting your walk back, you say goodbye for now to them and leaving the cabin you and your Higher Self walk back along the beautiful path in the woods.

When you are ready just come back to the moment taking some slow calm integrative breaths. This guided journey technique will help you to feel more relaxed, calm and certain about circumventing any unknown or unforeseen 'would be' events. You are safe now. You are sure now. Any life situations that you have asked for their assistance with have been intervened upon for you by the Fates, your Higher Self and you. You may proceed forward in life with more grace and ease and an internal knowing that All is Well.

Thank You, Dear Fates, for your help. Thank you Higher Self for being there with me. Coming back into the present moment, coming back into your physical self and grounding the energy in. Om Peace Amen.

Use this guided journey as your practice model. You may use this as needed for energy clearing and maintenance purposes. This is one meditation that you may not want to over use. But it's okay to use it as needed to get your practice down. After developing more of a connection with them and while using this same model, you can get more creative in your asking of the questions.

Chapter 4

PERSPECTIVES OF SPACE TIME REALITY

As we move through each moment we have unasked questions that are within us. Who are we, really? How did we get here? Why are we really here? Where is this all going? In this chapter we will cover aspects of space time reality and truths that exist behind it and sometimes just out of reach.

What is time? Time is a construct of the mind. It is a scientific, mathematical and yet creative delineation created for the intention of explaining the passing moments, the present moment and the moments which are coming that we have not yet experienced in our immediate and long range future. Our conscious human minds relate time with space as a way to organize events and locations in our lives and world..

The space time continuum reaches out across the event horizon and across the Universe coming from the originating contraction, compression and expansion of matter/anti-matter particles creating what the scientific communities call the Big Bang. Did the Big Bang happen in just this known Universe? Or did it happen across all other parallel and alternate Universes simultaneously? Was this Universe created after the other Universes (in the image of the Creator) as a newer alternate version of other Universes with modifications that are unique to our known Universe?

Time is relative to the perceiver. In our perception, the conscious awareness flowing through the individual minds, though all connected through group consciousness, the individual reference point at any given time interval can be malleable or pliable. It can become set in it's movement (contraction) or a flow of moments (expansion) occurring whether perceived individually or by a group consciousness. This allows for the relativity of time to seemingly move faster or to slow down in reference to conscious perception by us, and of spacial occurrences such as people, places, situations and objects.

An example would be that of someone rushing hastily to get to work on time as time seems to move forward too quickly. ('time disappears' or 'where did the time go?') The person then decides to consciously slow time down, or unconsciously by releasing all judgement of time constraints, and the trip is made effortlessly arriving them on time at their destination or even ahead of schedule.

Here is an interesting secret of the Universe. Time has consciousness. The world of quantum physics has arrived at the decision that All matter of the Universe, including all living beings and all inanimate objects are vibrating, some faster and some slower. And that Everything is 99.999 % empty space! Whether it's argued that the consciousness of time comes from our perception of it or that it is independent of us, could it be said that both are true? Time has consciousness just like matter. This is why it can flow at different rates, forwards, backwards or overlapping. So time as well as matter also has vibration.

A classic example of this is the depiction of a ball of yarn. All of the yarn strands are wrapped around and across each other creating overlapping intersections. This is why we can revert back into our child self or a past trauma version of self whether fully embodied or if only subconsciously. Which is to say, that our subconscious mind, which is in constant contact with the Universe can revert back to those times and manifest in it's present surrounding reality different versions of the same past events. It would not have to be only challenging events. It can also be good or memorable events based on what the subconscious is triggered to go back in time to reference. Indifference may show up as an uneventful normal everyday existence. Good events may show up as repeated progress patterns such as growth, love or abundant situations.

When someone wants to manifest something in their life or to create something, often times they can create a time looping effect in a way that what they are intending to focus on can always be in their future. Maybe they are not working with or writing/speaking/visualizing their affirmations for the occurrence to happen now or to have already happened. Possibly they are just not able to break through their own subconscious blocks of being uncertain or unworthy. This can keep desired goals just out of reach or it can bring a different version of the desired effect other than what they are expecting. Often times we can have or create subconscious games with ourselves placing the event we want to occur just out of reach. The desired event or outcome can even be in a person's energy field already present and they are not ready on some level to bring it into the 3D reality yet.

Let's say that it could be around personal healing, trying to heal something physically or a mental state of mind that is holding the person back. It could be around work or finances or a new relationship. When the person is able to gain ground by realizing the outcome which can also include letting go of the outcome, it then can materialize in their life. As the classic song goes: "Hold on loosely, but don't let go, if you cling too tightly, you're gonna lose control."

When the materialization of the desired or required event or object is realized in real space time, what a person has also done is to collapse time and space and to allow what already existed in the quantum fields to become localized and to become 'frozen in time' by being brought into our 3D reality for that person to experience or enjoy. It's also possible that people, through dysfunctional thoughts and emotions, or even fearful subconscious programs running in the background, could unconsciously activate Universal energy and create undesirable scenarios, situations or objects in a persons life. The more we know this, it becomes very important as to what you allow yourself to focus on mentally and emotionally. If you're going to collapse space and time and create an event in your life, make sure it's a desirable one.

As the Universe was created, it also has the built-in 'fail safe' self fulfilling mechanism of cause and effect or 'karma.' The law of karma states that for every action, thought, word and deed there is an equal and opposite reaction. This sounds like it could be daunting. If we were to take this word for word as law, then every positive action could have a negative reaction. Not necessarily so. It would be more correct to think of each and every action having a parallel reaction. Good generates more of the same and negative also generates more of the same. But what about the intervention of karmas, such as forgiveness, mantra cleansing work and gratitude? When we are truly thankful for a lesson, we are then seeing it for what it is and in so doing, released from it's grip and often gifted with serendipitous parallel reactions from the Universal mirror, which is interactively alive and in step with us.

The Universe has a built-in mirror system. As within, so without. So whether it's stars, galaxies, solar systems or planets right down to civilizations, countries, groups and individuals, the mirror system is always at work. It mirrors back to us what we really think and feel about the world outside of us. Regardless of our ego's opinions and finger pointing at what he, she, or they did.

Spacial perspectives of length, width and height are always at work in our minds and consciousness with our perception of the external world. Even unconsciously and subconsciously as we navigate life and our moment to moment pathway through it. Time and space are both constructs of the individual and group mind. Time and space are both 4th dimensional and accessed as part of our mind.

Our pineal gland in the center of the brain, often referred to as the master gland, is resonating the crown chakra atop our head. While the pineal is regulating all other endocrine glands in the body that is coinciding with each primary chakra, it has access to not only this reality but to all dimensions. The crown chakra is the gateway to the other incremental higher more subtle dimensions beyond physical reality. This allows for your angels and guides to stream in and connect with you as well as being a portal for your Soul's presence and it's connection up into the Akashic Records and into your Book of Life.

The crown chakra is connected into the Eternal Now Moment forever streaming oneness with all creation. For many, this is a very subtle connection and remembrance. For others it can be more pronounced. When you are ready through balance, connection, and raising your frequency vibrations you can open your crown chakra fully. This can help you to have a more tangible experience with the Universal Source Light of Creation, vivid Akashic Records journeys, visions, messages from your Higher Self and also a connection all the way back to the originating Source Light of all Creation. Back to the Universal Heart and Mind of the Collective Soul of God. This is when you are able to side step linear time while still aware of it's function. You are able to exist completely and continually in the Now Moment and stream it. This is when you are in the world and not of the world.

This can be Enlightenment for many and healing for others. You are able to see into and through the continuing moments past the confines of linear time. When you are able to be completely present within this perspective and continue to stream this, it will bring states of Enlightenment and Self Realization. You become self aware and aware of the Great Self. It can be healing as you let go of all regrets of the past and fears or anticipations of the future as you clear out, let go of and release the traumas that have been stored in your chakras, mental body, emotional body and physical body tissue.

Now this can also help you to have a more consciously aware and functional use of past, present and future linear time. When you can open to the Eternal Now and are fully present in the Now, you will find time disappears. It is no longer needed. You will stream life from moment to moment effortlessly and flow with the stream of life. This is what the Tao Te Ching calls 'living in the Tao' or 'the River of Life.' Some will have a profound experience of opening and then come back from it to reintegrate. Others will be able to cross over into the Eternal Now and stream it from 'Now on.' In ancient cultures this was a way of life for the adepts. In modern day it has become increasingly more difficult to practice or achieve this focus of awareness, and even though with all of life's challenges it is not impossible and quite achievable. The enigmatic mirror of the Universe and reality starts reflecting back to you a different variation of reality as it interactively reflects your inner experience in new inspiring and colorful ways.

This is one way to achieve a higher state of being that can lend itself to any other practices you are working with. This reality is a great holographic movie set that we are in. When you become aware of the dream while in it, you are now lucid dreaming. This gives you control and clearer perspective of everything you are experiencing, interacting with and creating. Including a greater sense of peace and clarity.

E = mc²
E = m
2
$R_{\mu\nu} - \frac{1}{2} R g_{\mu\nu} = \frac{8\pi G}{c^4} T_{\mu\nu}$

Chapter 5

UNDERSTANDING REALITIES WITHIN THE SUPER UNIVERSAL

Our lives are relatively small in comparison to our Universe and even our planet. And yet we are beings of Great Divine Power. This divinity lies dormant and hidden within all of us. Our great potential for mastery of time space reality, the elements, longevity, manifestation and above all the capacity to experience the Great Love.

This Great Divine Power is literally the consciousness of what we would consider to be God. Our minds and hearts have great capacities to process at ever new heights of expansion. And yet we tend towards our own expression of separation from it. We may often become inspired and have moments of direct knowing. These are indications that there is a greater life path and a greater reflection of the world for us to experience. As soon as we have these fleeting moments we often return to the underlying protocols or conditioning that we have either knowingly or unknowingly agreed to consciously, subconsciously or even unconsciously.

So how can we break these cycles and patterns that keep us spiralling from one drama to the next in life? We simply have to make the decision to change the script of our own reality TV show. No matter how overwhelming life may have become, there is always a way to rewrite the script.

Understanding that there are multiple realities in which you are participating is the first step. We have been programmed through the generations of religion to accept or be indoctrinated into the concept of one lifetime as opposed to many through the principle of reincarnation.

Does a Soul have to come back more than once? No, but unless you're a great enlightened Soul chances are you'll be here at least several times if not several hundred or anywhere in between. Souls who crash and burn or literally aren't 'getting it' are pulled out of the game. These can be repeated suicides, fatal accidents, life threatening illnesses, etc.

Sometimes this is when they will introduce a 'Walk-In' Soul to finish playing the Earth character role in the movie. Often this is done so because there may be one key Soul mate, other family members or associates that were supposed to get something of major importance to their Soul learning from the interaction with that Soul. So it would be important to keep that personality in the game. It really depends on what each Soul decided on for their learning. This is when the Keepers are doing major

Soul Tapestry work with multiple Soul players and their intersections of time.

The challenge is that every time we come back our memory is wiped before coming back, so we are working blindly through the ego, emotions, conscious mind and nervous system with whatever DNA programming we have from the lineage we were born into. As then we are given new and different scenarios along the same themes for Soul learning and observation. All the while the Soul knows everything. That's why it's so important to connect with the Soul. As we start to 'wake up' we are also able to access the DNA of the Soul body bringing celestial influence into our journey.

The one challenge we are actively participating in and unaware of, until we become aware, is the potential of creating more unintended karma. So let's say you are back here to clean up the karma of something that you really screwed up a few lifetimes ago. Remember when you burned the whole village and everyone in it?? (*example*) So you've spent the last 3 lifetimes cleaning up the karma of that event with those Souls on the Earth plane. And then...another ensnaring karmic event comes out of nowhere and you're in it. Do not fear, it could be a test to see how you react emotionally. Stay centered and as calm as you can be and walk across the fire. Remember in the end, everything will truly be okay. Better than okay, it will be just fine!

Hurray for those of us who can awaken to our remembering through the Akashic Records. That is the beginning of a quicker way off the 'Wheel of Life', as the Tibetan monks call it. Also known as a circular reincarnation pattern, cycle or loop. The paradox is that so many people here, even those in spiritual circles put a great deal of thought and effort into breaking the cycle of incarnation, while when we are in Heaven between lifetimes we know full well why we are here. Because we daringly agreed to come down, participate and do the work. We agreed to this, we are not being punished. So remembering this while you are here is a big first step and a great reason for accessing the Records.

Two key dynamic perspectives to be shared here.

1. Your Soul is connected with your physical/ Self here on the Earth plane and in the Akashic Records simultaneously.

2. Your Heart and Mind is also inter-connected with your Physical & Energetic Selves, the Universe and the Group Consciousness.

Both of these integrations are important and both can bring Self Realization and possible Enlightenment even if incrementally.

Many people who were staunch in their beliefs that there is no such thing as reincarnation have been jolted when they have suddenly witnessed a vision or dream in which they were playing out another lifetime. Psychology when paired with the lense of conventional religious belief systems can be dangerously limiting to a budding Soul's journey. I'm not suggesting psychology is bad in any way. It has it's place in our world. I have worked with many people as private clients over the coarse of my career, as an intuitive healer, who have been misguided, or held back in some ways because psychological protocols were being used as a designation for the effects and not the true causes, which can be metaphysical in nature. In other cases it has worked very well hand in hand with Soul healing and intuitive counselling. It could be that if you've studied or are actively using psychology, that this book and material can bring a new scope of helpful influence to your work.

Many people have also been taken to the Great Library of the Akasha for which they had no explanation. I've chatted with many people who have asked "What are the Akashic Records?" Only to reply once they received an explanation, "Yes, I've been there." Many who did not believe in the afterlife including atheists have come back from beyond the veil after a near death experience (NDE) and were shocked by the reality of what they entered into, whether their belief system was changed as a result or not. My question would be 'How could it not be?'

There are layered frequency based dimensions above or just outside of our waking state reality. These dimensions above or beyond our 3rd dimensional reality respond energetically much differently than our base reality here. The 4th Dimension being that primarily of mind. So one might say that when you are accessing pure thought or mental psychic gifts such as clairvoyance, extra sensory perception (ESP), telepathy, teleportation or telekinesis that you are accessing or employing 4th dimensional reality aspects. The use of these gifts however crosses over into 5th dimensional reality or what is commonly known as the astral plane. As then we witness the effects of these gifts sometimes in physical reality.

The 5th dimension is a whole other world in and of itself. You will probably have a more lucid or tangible experience with 5D rather than 4D. The 5th dimension is more spatial and contains expansive worlds upon worlds of different variations. Some will even say that this is where the hell realm exists. Not to worry. If you access the Astral plane you are not going to hell or anywhere near it. It is partitioned off and remote for good reason.

When Jesus said "My Father's house has many mansions" He was speaking about the 5th dimension or astral plane. It could also be said that he was speaking about the great number of dimensions that exist all the way back to God or the Central Source Light of All Existence.

The astral plane has complete worldly realities set up for someone that wants to go back in and complete an aspect of their life or even family life that needs more attention. Sort of like the Holodeck on Star Trek. A person with their personality/ego and Soul is able to go right in and be granted a certain lifetime of experiences. Their personal scenarios can be played out in linear time in these experiences. Other experiences may be a sequence of different events with different people which are strung together like different chapters or acts in a divine play. It may not take a whole lifetime length movie, but that is available, if needed. When a Soul, through their personality, is complete it can move up and back into the

Akashic Records where they will be assigned to another life incarnation on the Earth plane in this world, another world or perhaps in another reality. They may also choose to stay in Heaven indefinitely. It's a matter of decision for an evolving Soul. For the younger levels of evolving souls they are encouraged in coming back to Earth school. They are still learning the 'ins & outs' of relationships, right & wrong, love & hate, functional & dysfunctional, etc. and the Soul learns and observes through all experiences.

When we go through truly challenging situations on the Earth plane, we gather wisdom through experience. Whether it's health related, relationships, crafts and creativity, moral values, achievement, emotional or mental crisis and so on. Younger evolving Souls need more lifetimes to learn the ropes as they come back again for improved karmic relationships and generally better experiences within their world.

Ultimately your number of lifetimes required is not designated by the level of your Soul's evolution. Some Souls are quick learners and come back to achieve great things while influencing our world with creativity, innovation or leadership roles. When a person is coming close to their end of Earth life transition they will start to see glimpses, visions or streaming movies into the astral plane or 5th dimension. They can also be connecting up into the Angelic Realm as 6th Dimension, Heaven or the Akashic Records. It really depends on how a person is connecting on the other side and what their destination through projection will be.

In the 6th Dimension or Angelic Realm the energy

operates in a spherical fashion though it is very expansive. This is to house the divine angelic frequencies emanating from it as they radiate out reaching us here on the Earth plane.

In old religious studies such as Catholicism it was believed that everyone had a guardian angel. Some Souls still operate by this standard as certain paradigms follow them from lifetime to lifetime. Anyone can have angel guides, native tribal guides, family members, Ascended Masters or even elemental guides. It really depends on what the individual's needs or who they have called in through the Soul.

The early Catholic church was partially responsible for the idea of assigning an individual guardian angel to each person. Maybe it was felt that in earlier times each person needed the protection of an angel through the dark ages. Let it be noted that in some ways they were reorganizing or simplifying concepts from other world religions. Such as that individuals have guides on the other side. And in other ways they were changing the foundational structure by largely governing the naturally evolving principles of the hierarchical religious worship and studies. It was the new improved version or model in an evolving world. While certain aspects, teachings and historical data were changed, 'borrowed' or 'improved' upon, other aspects, teachings and historical data were left out whether purposefully or possibly deemed secondarily unimportant in those current times.

It's like the scenario of a growing town that wants to become more modernized. As the new trends of society arise and time marches on, they commence to tear down half of their original historic buildings. Then later in the future, as they are refurbishing their historic area, they look back with regret on tearing down half of 'old town' only to replace it with the newer modern style of tin buildings and stucco store fronts.

For the church, something similar applies only on a grander scale of changing generations of world history information and spiritual studies. I truly think that this was the case in the reorganization of the Bible. This is a delicate topic for many. They thought, in that time that they were improving on and / or poetically embellishing with the whole rewrite primarily for the current modern Europe at that time. So things were changed or lost. It wasn't diabolical, it was, in their eyes, newly improved upon for a changing world and so more people can understand it. The true essence of God and the Holy Spirit which flows through all things is very present in the compilation so the vibration was never lost.

The Angelic Realm or 6th Dimension has governing or leading Arc Angels who each have legions or armies of angels for certain purposes. One thing is for sure the Arc Angels are beings of great power. In one way they are like the Ascended masters. They have unique functions and unique personalities. The Arc Angels do powerful things in the higher dimensions and on the Earth plane. They influence humanity in a multitude of ways. There are some that believe that the Arc Angels are of extraterrestrial nature. I think that some of them are and work with the Galactic Federation of Planets and the Ashtar Command. The frequency that vibrates from each of them is unique and radiantly powerful. Author Dr. David Hawkins calibrated world teachers with the 'Hawkins Scale' of vibration from 1-1000. He found certain avatars and teachers to be at 1000 such as Jesus, Buddha, Babaji and other world servants. He then calibrated the Arc Angels at 15,000! So again the Arc Angels are beings of great power. We will touch in with the Arc Angels more in a future chapter.

So Higher Souls and beings of great power can exist in the higher frequency dimensions. From the 6th dimension on up into the 7th to 12th dimensions. These have been regarded by ancient spiritual cultures as sacred heavens. Mostly unvisited by human incarnated Souls though it does happen. And then Souls make the jump up into the Akashic between the 10th, 11th and 12th dimensions. There are Great Souls in these regions that have never incarnated into a physical reality here on Earth or any other world.

While the true number of possible dimensions is unknowable by means of human consciousness, we have theorized many numbers of dimensions. Ancient Tibetan and Asian cultures believed in 7 heavens. Some ancient cultures believed in an infinite number of dimensions. Modern day quantum science believes that there are 11 dimensional realities. Some believe that there are 33 dimensions. Other religious groups through history believed that there were 72 or 144 dimensions.

In my life's work of accessing the Akashic Records I have learned that there are at least 12 dimensions. They have shown me that the true Akashic Realms are located between the 10th and 11th dimensions and then again between the 11th and 12th dimensions. So while less Souls are visiting the 7th to 10th dimensions, most all Souls participating in the incarnation process visit the Akashic Soul Temples in these higher locations between the 10th, 11th and 12th dimensions as part of Heaven and as we Souls know it to be. The Akashic Realm is directly connected to Heaven as we know it.

This is a heavenly and heavily guarded place. This is God's Library of Souls. All Souls coming to the Earth to start their incarnations enter here through the gateway or 'customs' to being born on Earth. Reincarnating Souls who are on Earth and do not make it back to the Akashic Hall for reassignment can be due to being emotionally lost and wandering in the astral plane from suicide or other Soul separating traumas, until they are possibly redirected back into the light or rescued. Some Souls become lost to the hell realm. Others can be transported to other worlds by means of extraterrestrials at which point they may regain their point of origin or begin reincarnating in another star system. We Souls sometimes willingly graduate from Earth School and return to our originating homes among the stars or to new worlds that we would like to experience.

The process of incarnation and reincarnation by the individuated Souls of God is a multi-world program, each world having it's own Akashic Records program. This is set up to evolve the Souls. The Souls become greater and greater beings over the great expanse of time in this

Universe. You might think of it as God evolving us as Souls so that eventually He has a friend to talk to on His level. There are a few higher level Souls that are getting close to that level as Great Beings of Light.

Let's get back to time-space reality for a moment. There are an infinite number of realities. The existence of Infinity has been proven through science. There are different realities or Alternate Parallel Realities where we are playing out a multitude of different life variations. A Russian quantum physicist by the name of Vedeem Zealand has called this the variants space or the alternative space. From these other realities we can call into our life movie an improved version of life or a new event to happen as well as achievements and acquisition of things desired simply by choosing.

This is a new concept from a quantum point of view. Most of the past and current schools of thought are focusing on the manifesting of that which is focused on or felt as an emotional response while imagining (or imaging) the outcome from this reality. This is often thought of by the individuals who practice manifesting as bringing things from the seen or the unseen of this one reality. By simply focusing more on bringing it here into this reality from another reality where it already exists is another unique way. Both ways will work.

Our core or intended reality is surrounded by these alternate realities which travel through time and space forward. Think of a space-time tunnel with many space-time tunnels around it moving through Eternity. It is possible for one of these realities to bend and dip down into our core reality for a short meeting or to become ingrained over a longer period of time. This can influence people places and things, the whole reality shifts in a seamless way that can be silently unnoticeable when it occurs through bringing about anything from subtle to quite pronounced and noticeable shifts and changes.

As people look around scratching their heads saying "What is happening to our society these days?" This reality overlay can be for the better or it can create changes of decline. Or it can simply be different. Usually the group consciousness of humanity along with other beings can create these shifts so it's usually for a purpose and not indifferent.

It can happen for a whole planet, a whole civilization, a whole country, a whole race, a whole group consciousness within any of these, a whole political party, a whole group or family or simply for just one individual. Not only this, but it's happening on all of these levels sometimes consciously out of intentional shifts and sometimes haphazardly to any or all of these groupings. When you study reality on this level you become more careful about what you think and feel about. What you intend or choose in your reality. Choose Love. Love always works. Love conquers All. The truth is out there. And the truth is also inside of You. And...

The Truth Will Set You Free.

Chapter 6

Conscious Perception and Awareness

Our perception of consciousness and awareness is processed and experienced through our senses via the mind. The consciousness of the Universe finds us before it is translated via the mind and senses by way of our Soul field and multi-layered auric field. We find the sense of perspective, depth, light, dark and colors through the eyes. Sounds through the auditory system, smells through the olfactory system and touch through the nervous system.

From these base operating systems we are also able to sense on other levels. We can sense certain energy shifts in our surroundings via the empathic field surrounding the body, the higher functions of our mind and the nervous system. We can sense higher spiritual vibrations with our Soul energy as it radiates into our being and can influence the mind, the senses, DNA and all tissue. We can also sense a multitude of vibrations in our surroundings via the chakra system in our spinal column and our intricate nervous system.

We continually utilize perception through thought with our mind field via the brain. Here in the arena of mind we process linear time as past, present and future. We think about life and our personal place within it. We ruminate on regrets of the past and fears of the future through the awareness of our ego. This is also giving rise to the consciousness of time.

As we re-think of memories or project forward in time we create or reactivate a direct interface with those personal moments of past, present and future possibilities bringing them to the 'Here and Now.' Reactivating those moments through the mind and nervous system as the subconscious mind is listening. Will the subconscious take the bait and act upon these projections? More so, if there is an emotional charge associated with the memory.

As Neville Goddard has said in his early works on manifesting, the subconscious is like a mistress or wife. She may act upon the thoughts and feelings creating new events in our reality whether positive, negative or indifferent, or new variations based around an old theme. Though, she may also choose not to act on these based on indifference to the situation as she deems fit or necessary (through the nervous system).

This occurs when we are focusing on being positive and in gratitude, rehearsing our vision, saying our prayers and affirmations and nothing is happening, even if we think we are 'in the zone.' This is when it may be helpful to interrupt the process of life with relaxation, exercise, taking our minds off of things and then restarting with newly found inspiration. You may need to take things up a notch by making new changes, becoming more vocally passionate about your desires or moving the physical body which can clear and realign the energy for you. For instance, if you are doing mantras laboriously everyday and nothing is happening, look at the quality of thoughts and emotions you are in while you are doing them. This can change things as the subconscious and the Universe notice the shift. Sometimes we need to consciously interrupt patterns in order to get out of our own way.

We search for meaning and process creativity in all of it's many forms. We also process a wide range of emotions with our emotional body. So our awareness even on a base level can be very astonishing when we look at it more clearly. You may use the idea of being the 'Watcher on the Hill' in meditation or thought to take a look at what you are doing in life. How are you reacting or responding to the world around you? What choices and decisions are you making. What are you truly focusing on?

Think about all these sensual systems operating simultaneously. The chakras in the spine are connected to our multi-layered field surrounding the body. The etheric, subtle, causal, mental emotional, astral and etheric bodies are the layers of this field. We are sensing with the pineal gland in our brains all dimensions though we are not aware of most of it. We may access other dimensions beyond 3D through dreams, meditation practices, fleeting moments of enlightenment or even an emotional epiphany. We sense clairvoyantly and telepathically with our 3rd eye. We also sense mystically with our heart. The heart has all the answers and what it really wishes for our best interests. Though it takes a back seat to the mind which is assuring the heart 'I'm the one driving the car and I know the best way to get there.' After hours of searching and getting lost, the heart quietly says, 'I was trying to tell you, I have a map in the glove box.'

How do we explain a sense of separateness with each other and God/Creator/Universe and yet still have a sense of Oneness and how can both seem to be true?

Our first sense of separation at the base level comes from our individual bodies which are all unique in shape sizes, complexions, and personalities. Our egos then through the base operation of our nervous system, lower brain functions and senses have a primal operating system based around survival, fear and competition.

The ego's original functions for conflicts in a primal world through the unconscious primal self have been:

Man against man *(Self against another self)*
Man against nature *(Self against mother nature)*
Man against himself *(Self against self)*

For today's modern world we also include:

Man against the world *(Self against the world)*
Man against God *(Self against God Self)*
Man against woman *(Masculine self against feminine self)*
Woman against man *(Feminine self against masculine self)*

The ego is crafty, cunning, silent, outspoken and can operate in a myriad of ways to get what it wants. In today's world of projections, the ego has new versions of conflict. It now has new standards to keep up with such as: who is the best in their field, keeping up with the latest fashions, the newest model of car, a bigger and better home, managing and / or limiting relationships and interactions with others, political parties, sports teams, world views and the list goes on. The ego can also keep us separate from our spiritual connection of inner peace, away from our Higher Self or away from our God / Source connection whether consciously or unconsciously.

Other levels of separateness that are perceived through the physical world are as follows. Everything thing in our world is of differing shapes, sizes, color, textures and quantity. We specify the differentiations of all things. A table is not a chair and vice versa.

Directions: left and right, reverse and forward. Polar opposites: dark and light. Temperature: warm and cool, hot and cold, scorching and freezing. These are often thought of as opposites instead of options. 'It's a big world out there, you don't want to get lost, so be careful.': This is separation. (ego's perception lense of impending conflicts)

Our thoughts and emotions are generated from within, or so we think. Thoughts are more rarely unique than you would think and emotions, while seemingly being generated from within, both thought and emotion resonate with bands of frequency in our surrounding atmosphere that we simultaneously tap into.

In another light our individuality is important to us. Our unique gifts and qualities are recognized by ourselves, the outside world and others. It is simply the dynamic of the way we show up in this reality. Though we move together in schools or herds of thought. We group together in schools and universities, sports teams, religious and spiritual gatherings and meetings, clubs, villages, towns, cities and states, countries, continents and civilizations. The truth is that we are not all equal though we tend to try and equalize everyone into the unity principal of grouping. A stained glass window has greater and lesser pieces of glass of all shapes, sizes and colors. They are not equal, though each and every one is needed to complete the window.

Most of our ideas and views of Unity and Separation are subconscious or on auto-pilot. There is a Great Field which houses all dimensions and realities in the Universe. This is the ideation of Oneness or Unity. Quantum physics calls it the Quantum field or the zero point field. We are all connected through varying levels of group consciousness.

Group consciousness can be influenced by thought and emotion. This is where we tie back into the principle that less of the thoughts and emotions are more original than you think. Thoughts occur continuously at different intervals of time making them seem personal. In society we tend to celebrate over thinkers but what about the under thinkers? The great masters have said that less thoughts are more powerful. In that, when you do have a thought there is more energy focused with it, due to there being no continuous chatter in the background of your brain and mind. Think about it.

The Originating Source Light of Creator God is

comprised of the Collective Soul of the Universe. It also emanates the radiating light particles that are building blocks for stars and celestial bodies, suns, planets, beings of light and worlds upon worlds (dimensions). This Light also carries the signal of what we know as Love. Creative Light that creates the expanding Universe and Love are one in the same. The light carries the frequency of unconditional love or Creation / Creator's Love.

This level of love is often beyond the understanding of humanity. We may say that we understand it though our thoughts, words and deeds may show us otherwise. We can generally be very judgemental beings. We do understand the idea of Unconditional Love. And in some ways we are able to channel and interpret this loving, healing and creative power. Though we find that most of us are not able to continuously express unconditional love and we have to practice at it. This is due to societal programming through separation. So, practice and don't give up on the opportunity to express Unconditional Love in every aspect of your life.

It's here, there and everywhere present just availably waiting for that fleeting moment when it shines through us in ways that we can feel and understand on whatever level we can allow it to flow through us. This has been called in history through spiritual writings, The Presence of God, or the Holy Spirit, and so many other creative expressions. This Divine Presence is what is experienced when people have Near Death Experiences (NDE) which more accurately could be called a Near Life Experience.

The Collective Soul births individuated Souls that go

out from it and seek to enjoin with physical life forms to learn, observe and evolve.

Going out first as each Monad or Over Over Souls
Then out from each of those to 12 Over Souls
Then out from each of those to 12 Soul Groups

These Souls span the Universe looking for intelligent life to blend with. When this Celestial Soul blends with the body Soul of a Life being, it incarnates together hoping for a harmonic alignment with that being. Once the right blending is achieved, the now newly enjoined celestial Soul carries the frequency of The Originating Collective Soul of Creator. This is the evolutionary process of higher consciousness within civilizations. And it's been going on for much longer than we know. This Soul enjoining process was originally known and even practiced in spiritual circles and rituals and then over time it was forgotten, lost and even purposefully buried. The Universe is much older than we know. With every new study that comes out the possible time length of history increases.

So we experience a sense of Oneness through this Soul Matrix. We can often find other Souls in our Soul Group or even extended Soul family and feel a sense of kinship or partnership. Many great things have been achieved through these connections. World movements, innovations, personal or collective healing, right down to personal relationships whether intimate,friendly or transformational.

In summary of all of these variations of separation and oneness or unity, we each also have this dynamic present and operating through us:

Our own ego or lower intellect is operating
In Us, As Us and Through Us.
Our Celestial Soul or Higher Self is operating
In Us, As Us and Through Us.
And Creator God's Presence/Universe is operating
In Us, As Us and Through Us.

All three of these overlays are present in every person's consciousness while either silent or realized. And most of us shift back and forth from one consciously aware perspective to another, whether we are conscious of it or not. For the newly beginning incarnates, they are primarily operating through primal ego while awake and then connected to Higher Self or Source while asleep. For the intermediate levels of incarnates, they are becoming more aware of the shifts happening and learning to move back and forth recognizing the overlays of consciousness and learning to be present to more than one overlay simultaneously. For the more advanced evolving Souls, they are learning to heal the ego and utilize it's positive aspects while streaming the Higher Self and incrementally streaming God's Presence until they can fully embody it leading to enlightenment and Self Realization. These are generalizations that occur and shifts that can be realized for an individual at any level.

We are now seeing newly incarnate Souls coming into the past few generations as being more awake. These were originally called the 'Indigo Children.' Now they are being called 'Star Seeds.' As these pockets of new Souls are coming in with heightened awareness and abilities, they will be influencing our world and the coming generations in new positive ways. While this is happening, we still have a tremendous amount of Souls incarnate still doing it the old school way. Though, we are waking up slowly and incrementally through each of our own life path processes in the grand scheme of things together as a whole and individually.

So be vigilant Grasshopper, to be on the Path you must first become the Path, your Time of Realization and Enlightenment is at hand.

Chapter 7

Delegates from Beyond the Veil

On our journey through life, as we awaken or reawaken to the world of Spirit, we may find that there are a great many teachers, healers, helpers and guides available to us for counsel, messages, healing and help on our Life's journey including manifesting the things and experiences that help us move forward to help improve our response to life and the world around us.

These Great Souls have come to be by way of incarnating on the Earth or other worlds and holding their light through learning, life experiences and service. As a result they evolved whether through a series of incarnations or from achieving a level of enlightenment housed within their being that generated frequency alignments.

We refer to them commonly as the Ascended Masters. Each one of them with their personality intact from the lifetime or moments of great awakening that they experienced. They have ascended the wheel of birth, rebirth and death as we know it. Each one has achieved unique gifts from their own spiritual practices. Their diversity is gathered from a multitude of practices, experiences, service work, devotion and study that has given each one of them similar powers. Referred to in historic Indian spiritual studies as the 'siddhis' or spiritual powers. Even so, along with these siddhis they each have unique abilities they use in their own personality's ways. Not all masters with powers were practicing these Indian techniques. They can be acquired through other forms of practice, service and study. So if you are working with Quan Yin her presence, messages, manifestations and healing can be different from that of Saint Germain, who works with the Violet Flame and the Holy I Am Presence. Jesus will work differently than Buddha and so on. Another dynamic worthy of mention is that the Ascended Masters can work much differently after their ascension than they may have on the Earth plane. As they can now do even more and in different ways.

The Ascended Masters are a coalition of primarily human Souls (along with some other worldly ascended personalities) that have evolved or ascended. Once ascended, they can remain in each of their personalities indefinitely. They are immortal by our standards in a suspended state of animation. This is available to all incarnate humans, we just have to be willing to do the work in that direction. All Souls enjoined in a human body have the possibility of finding their way onto the Great Path towards ascension. As they do, the Soul enjoined with that physical body awakens the body's dormant mystical abilities. This has been explained through ancient spiritual texts and in modern day spiritual and ritualistic circles.

Most commonly when people talk about energy they refer to the chakra system. But there are many other elements to the physiological makeup of humans that make us potentially mystical beings. The Brain, the nervous system, the DNA, the endocrine system, the blood and the organs all have potential mystical qualities. We project energy from all of these parts of ourselves when we pray or meditate and actually through everything we think, feel, say and do.

It has been stated through ancient texts and discussed in modern metaphysical/scientific studies that we were genetically modified and engineered to be like the 'gods' (aliens - specifically the Annunaki) as workers to populate the planet and serve these visiting 'gods' in mining for necessary minerals (primarily gold) needed for space travel, and to help suspend the atmosphere of their planet. Even so, if they genetically modified Homo erectus and then bred early versions of them to quickly multiply, the Universal Creator had a Divine silent hand in our creation. The late researcher and author Zecharia Sitchin in his Earth Chronicles series believed that the Nefelim were the offspring of the Annunaki. As the Annunaki bred with these early genetically modified humans. Then their offspring, the Nefelim bred with the Earth humans that were starting to live in small tribes, and over time modern day Homo sapiens were the result. And this is how we learned to war - from the gods(aliens). They had a distinct trait for superiority. How could they have known that they were also creating a body type capable of ascension when it would be paired with a Celestial Soul? This is one characteristic of humanity that is miraculous in it's own contemplation. We are, just by being here as Celestial Souls enjoined with human body and channelling the Universal Consciousness of Creation, a living miracle.

The graduating class of Ascending Souls is small in comparison to what it could be. So many Souls tend to come down to the Earth plane with good intentions and inadvertently make varied choices and actions which can then bind them to a greater amount of incarnations than originally planned or needed. This happens of their own choices, but is also heavily influenced by karmas of the bloodline they incarnate into. Eventually we find our way. Self Realization, Enlightenment and Ascension are always potentially available to us within each moment of each lifetime, though it could take more than one, several or many lifetimes to achieve it. Even after that we tend to come back and experience different variations of learning along familiar and different threads.

Our great challenge is the loss of memory that we experience each time we are reborn into a physical body. Generally in modern generations we naturally find our way back to our inner knowing or Soul's vibrational creative purpose either by intending to search for a higher meaning, or it finds us at a certain moment and we awaken or re-awaken.

Our other great stumbling block is that the Earth plane is full of Souls who are all evolving at different levels and speeds. While in our world we tend to either unite under equality based on physical principles that make sense or we tend to separate into groups which are different variations from others also based on physical Earth plane principles of existence.

While almost None (very few) recognize the different levels of evolving Souls among us. We mainly recognize our differences through accomplishments or failures, wealth or poverty, privileged or under privileged and so is the world of duality. If we were able to see these Soul levels more readily in a spiritual sense, the world would be a different place. Like realizing the difference between a child, teenager or adult physical human, realizing the difference between multiple stages of different Soul levels. We generally recognize the Great Souls by what they can do or have done and yet there are many among us living a normal life silently, helping neighbors, teaching school, having families and so on.

If humanity had evolved more consistently from some of the more advanced previous versions, this piece of the puzzle alone would have been one of several harmonically aligned group traits which would have allowed for a more creatively unique and unified version of civilization. A society in which the evolved Souls were recognized for their ability to contribute to the evolution of all other Souls. This does happen in our world but were are lost to it's operations from behind the scenes.

Atlantis had this. Ancient Greece had this. Egypt had this. And other spotted kingdoms over the course of time embodied this principle as their royal family, priests, mystical school circles, shamans and wise men were all as servants to the whole community. The renaissance period of Europe with Vienna, Austria at it's center, was leaning towards this through the arts and music culture. This has been experienced in different ages, by different peoples around the globe. Atlantis, Lemuria, Native tribes, East Indian culture, Asian communities, and others implemented these designations of advanced Souls, intermediate Souls and growing, learning Souls. Places and periods when it felt good, creative and comfortable to be alive. And then there were those rulers who controlled through duality and separation in the same and other historical ages. Another representation of separation and unity.

Whether due to Earth changes, neighboring conquerors or imposed changes due to rulers who thought they knew better for the world, a higher constant way was not able to be sustained. Though, throughout time, the dynamics of harmonic alignment with higher frequencies of 'all that is good' keeps showing up, forever present, through events, groups, individuals and yet sporadically through history. We struggle to make peace with it, explain it away and learn to live with it as light and dark, good and evil or 'duality.' Never able to maintain a constant good run long enough to create a lasting shift.

Until, of course, we make the great Cosmic Shift back into Unity. Our world civilization's growing pains have been seemingly long and drawn out.

The Ascended Masters are those evolved Souls who have gone before us throughout these sorted times. These spiritual human examples have shown us repeatedly what is possible for all. This is why they shown so brightly in contrast to the world around them. It was almost as if they were from another time. They are aware of the shift that they have made. They are able to visit the Earth plane as well as shift between the other dimensions. They remain present to help awaken those who can be awakened to start their Soul Evolution path towards the journey into Ascension. They are way showers. There are principles that some must abide by. Certain Masters have clearance to affect the Earth plane. Jesus is one of them. Mother Mary is another and Merlin is yet another. It depends on what we are asking for and what their agreements are with the Universe.

As guides they can be quite healing, protective and revealing to greater truths and our own personal truths. Your interaction with them also depends on which ascended personality you are connecting with. From time to time we may ask a certain Ascended being for something that either they are not willing to do or is outside of their scope of being able to provide. They can 'nudge' the energies of physical existence though in ways that can equal big changes down here.

Generally, it's always best if we can try to accomplish things on our own. That is part of the Earth School training/learning program that they had to go through to get where they are. They are guides, way showers, teachers and message bringers. It is possible for them to manifest for us, usually by healing and clearing, though, after spending time with them and learning, when we no longer rely so heavily on their influence we are able to step more fully into our own power, while still recognizing theirs. This helps them to help us in even newfound ways that we couldn't even conceive before. You have the power through Knowledge, Wisdom, Study, Practice and Love.

Things just start happening for us. We become more in step with the Universal energy of time and space. We have the Universal principles of energy to work with and to learn how to use as we create through our own lense of consciousness into this reality. The key is to keep our priorities in alignment with a greater sense of purpose. Manifesting for the purposes of the ego's joyride is short lived and virtually pointless in the grand scheme of things. That's one reason among many for an extended amount of reincarnations. It's OK to manifest what you want. You will have a better track record if you manifest in alignment with a greater purpose for self and others that is beneficial to the collective in some way.

The angels and Arc Angels are also delegates to the Earth plane and it's civilization. Helping humanity to grow by guiding us in certain directions or along certain guide lines. While we do have free will, the Ascended Ones and the Angels can help to protect our path from outside influences or they can actually influence or even direct our path when needed.

Some Souls are guided by one Guardian Angel or one Arc Angel and it's decided on by the person and by the angels, and that's enough for that person or Soul. Other people may have certain multli-layered aspects to their creative Soul evolution which can attract the attention of more than one Arc Angel. They sometimes contend with each other over certain gifted Souls in an attempt to attract the attention of that Soul in their direction. Or we may be calling in multiple angelic personalities that bid to work with us.

The bidding and countering energies can be natural on their part as the Great Light from them is so powerful. When we lean in one direction or thought, emotion or life, we are working with one of them and when we lean in another direction or thought, emotion or life, we are working with another of them because of the shift within us. It's interesting for them to work with the higher evolving Souls, while they are bound by God to work with all Souls on all levels, all the time.

When a person is attracted to more than one Arc Angel sometimes this can present a bit of energetic carousel in the person's life. Remember they are beings of great power and radiance. You may want to limit your working with them to a fewer number of their ranks. Isn't more better? Not in this case. To some more evolved Souls it won't matter much because they may be obtuse to a deeper focus of outside influences in their development. For the energetically sensitive and aware type of Soul/person it can be overwhelming to get developmental gifts and lessons from more than one source that can be in natural varying degrees of opposition with each other's energy as it is translated down here an into the Earth plane.

Remember they all have unique qualities and personalities coupled with great power and radiance. Some of them get along better with each other and some are very unique or outspoken, creating seeming contrasts of agenda for a lower being (us) in a lower energetic arena of 3rd dimensional reality. One Arc Angel may be concerned with you being of service. Another may be concerned with you growing your abundance. Another may be concerned with your interaction with large groups of people. Another may be concerned with clearing, protection and justice. You might see how these elements can work together in a person's life or how they may be disruptive, slowing or opposing to us beyond their original intentions. This is one reason we have chakra systems, to filter and regulate how much we can take in. Remember in light of this, everything is okay and everything gets used.

Their vibration is high and they are on call with Creator God and Jesus for everyone. Unless a higher level of angels is dispatching a lower level of angels (their legions) to intervene with your path in some particular way. The demons will fight over you sometimes. Demons, like the reptilians, feed on lower vibrational thought and emotional energies. Anger, hatred, fear, disgust, jealously, woe, hopelessness, deep depression, loathing and the list goes on. They are attracted to these energies we give off to feed on them. They can also influence these energies to come about by intervening with our surroundings in some way. Lower vibrations of negative energy through thoughts and feelings are more tangible to them and they may be able to come into alignment with you long enough to steal this energy or even cause you to generate more of it. Light, Love, Compassion, Joy, Healing, Gratitude, Forgiveness and Service repel dark entities, reptilians and other negative aliens, and demons. They cannot stand it and cannot be around it.

Do they show up first and then the lower vibrations? Or do the lower vibrations come first and then they show up? Take your pick. When you focus on the Light, Love, Forgiveness, Gratitude in your heart, it drives them away. Sometimes you may need to call for the Arc Angels to intervene. "Hello Michael!" Michael has strength and power. If you're going to call on anyone to get something done in the clearing and protection department, Michael is a good first choice.

You can say a thought word prayer command such as: "Dear Arc Angel Michael, please come and clear any and all negative energies, entities or beings that are not of my higher being or for my highest good from my physical and energy bodies, my home and my life."

Keep in mind that they all have extremely high vibrations and their presence alone can thwart danger. I'm not going to focus too much on demons. They do have legions like the Arc Angels, but they are parasitic and vampire like towards Soul energy. Promising a person personal gain and power if they go in a certain direction. They're after control of a weakened or confused broken Soul, if they can break or fracture it and keep it that way, then they can get energy from it and maybe even a promotion from their boss or upper management and any other Souls they can gather as a result. The more the merrier. Remember this: pure evil is based around justification. How many things in the world have been made to be justifiable? Wars, greed, stealing, manipulation of the legal system, lust, giving up, violence, lying, etc. These are all versions of the group or individual ego's justification that something is alright to engage in.

For example, the dark ones may be courting some female singer who dresses provocatively and plays for a dark speed metal rock band. She may attract a great many Souls to her through young male and female concert goers. They often work through people by promising to give them energy such as fame, money or power to get to many other people. The thousands or millions of concert goers may be indulging in alcohol, drugs, toxic relationships, violence, crime or other lower vibrational activities. So the dark side's customer base just went up. Soul energy is a currency for these beings.

I have seen scenarios where dark spirits can groom and be in direct manipulation of multiple people in a family, multiple friends or business situations. Though, Souls can do this too. As well as the Arc Angels. One example of a great Soul influencing a group is Jesus and the Disciples. The Great Soul coming through Jesus over lighted the Souls of the top 12 disciples to carry out the story line which we know so well, (there exists a more original version of that story, but that is for another time).

Angels Save a Life

It was several years ago, that I was doing a workshop in Indianapolis, Indiana and was coming up the Eastern seaboard from North Carolina, through Washington DC, Virginia and then across rural Maryland, Pennsylvania, Virginia, Ohio and into Indianapolis, Indiana finally. We had a successful 3 day workshop at a large hotel there. I rested for a day and contacted a few local spiritual bookstores to see if they were interested in carrying my books. In northern Indianapolis, I found a very large unique store. As I pulled the RV travelling coach and my tow vehicle into the parking spaces adjacent to the roadway in front of the store, I went in and met the owner, visited and looked around. As we agreed on a quantity of certain book titles and CD's, I went out to the RV and gathered the books, CD's and promotional product. When I came back in and went to the office in the back of the store, the owner introduced me to a friend that always stopped in to visit with her about this time of day.

Upon receiving a check, I walked out into the store, on my way out, and decided to buy a few things. As I walked out to the RV and stood still in the RV before leaving, there was a slow motion moment of stillness and quiet that came over everything. A great hush. All of a sudden I was knocked to the floor by an impact of a huge crash. I went out to find 2 cars one behind the other had left the motorway and smashed directly into my Saturn tow car behind the RV demolishing it! So an hour later after the police had taken a report and the tow truck had arrived, I was standing in front of the store checking the RV over. My tow car had been loaded by a wrecking truck along with the other two cars and all three cars were now gone.

The owner walked out and said, "You know the strange thing is, My friend that I introduced you to always backs his truck into that spot where your car was and comes in to visit while his wife sits in the car on the passenger side and reads. Today he had to park in the back lot because your RV and car was taking up the parking in front of the store. If he had parked there today, his wife would have been seriously injured or gone!"

Immediately it occurred to me, that's what the amazing stillness or silence was, (just like in the movies) everything was too still and in slow motion. When you know something's about to happen. The Arc Angels had placed me and my travelling coach there at that time to use the RV as a blockade And to save a life! All the way across the country, with the timing of multiple workshops in several states to be in that very parking lot at that very time. This is a prime example of how the Arc Angels can use multiple people, places, times and events to create an outcome. I'm sure many of you reading this may have some interesting stories of your own!

The Celestial Souls

The Celestial Souls move out from the First Moment of the Collective Soul of Creator God Light Source of the Universe as Divine Delegates. Searching for life forms to merge or enjoin with. "I Am the Lord of Hosts" is the ancient message from the Holy Bible, the Gnostic Gospels and other sacred writings. The hosts are the life forms, that the Celestial Souls seek out to merge with, to evolve with and to influence the growth of the civilizations among the stars. We are those hosts. As the Souls enjoin with the body spirits of individuals whether at the birth of an incarnation or during an incarnation, the Soul brings a presence of the Divine Energy Light of Creation.

Something more that starts to evolve the human being. That being is You. This dynamic is experienced by individuals that I have worked with time and again. There is usually one person in the family tree that wakes up to a greater inner knowing often times without a reference point to some of the many thoughts, feelings and energy sensations all around them. It can create a rift between family or partners when these moments of realization occur. It's just a difference in the vibrational shift. They may say "Something is different about you, I can't quite put my finger on it." or conflicts may arise out of a primal need to stir the old family ego.

Some will try to stuff it down while, like trying to gather water, it flows between their fingers and surrounds their hands. Sometimes it does not create turmoil between the surrounding others. There is a greater purpose in motion, a great question that must be answered. In truth the question and the answer has already been given in the higher planes to that Soul/ incarnation/person.

This is the time in their dream (Earth life) when they will act on it or step into it and play it out. For many there can be conflicts with life in general that arise. Which life path decision do they make? Which family, partnership or friends will they surround themselves with? What gifts, accomplishments, victories, addictions, traumas, accidents or other life changing events will they play out? Many can experience finding their purpose, finding the right love partner, creating great art forms, experiencing abundance and success, a deeper sense of spirituality and spiritual learning or practice and leading a fulfilled life as a result of any and all of life's many colorations of experiences lived.

Can any challenging sequences of events be changed for a person? Yes. Through deeper self study, forgiveness, energy work, prayer, and a drive to learn the understandings of how the Universe works. You are doing it now by reading this book. "But how?" the ego asks, as it chases it's own tail around in a circle.

Through your Soul, the Higher Mind and the Heart. All of these parts of you have an inner knowing and the ability to observe beyond lower conflicted endless searching. The Soul, Higher Mind and the Heart have the ability to observe and as such, gather wisdom through direct participation or by being the 'Watcher on the Hill' with the external, the internal, previous life situations, childhood, past lives, social interactions and spiritual recognitions of the world. The dream in which we are creating, interacting with and co-participating in.

The ego is always going to be looking for the quick hack method or a magical boon that will make it feel special. In so doing it can try to manage a person's spiritual growth by spinning a person through an endless search when the Real Self already knows what to do. All of these things are helpful and as the person is always in an environment to make that inner shift. The question is, will they? Sometimes we can feel it coming, other times we may put up walls to block it knowingly or unknowingly, or we may not even see it coming. I'm talking about the shift into true Self Realization.

Manifesting and manifesting gurus are a good example of this (no negative spin intended). The ego wants to understand a process that you are already aware of on a higher level and that you are already doing. It seeks out a new improved version of getting what it wants. 'Then!...Then I can relax and feel like I've accomplished something!' It's completely Good to learn the dynamics of creating and manifesting and to achieve the things you want, need and desire in life. We are just referring here to the ego's self appointed position of clandestine or outright inner control of a person's creative spiritual process. The key is self study in this department and balance. Be the 'Watcher on the Hill' and 'keep your eye on the ball.' More an the ego later.

All Things are God

The First Moment, Divine Center of Creation continues timelessly broadcasting out the Cosmic Light of Creation. These ceaseless light particles hold the pure Essence of Creation as they move out creating stars and suns, celestial bodies, galaxies, solar systems, planets and life on all those planets. The Cosmic Light particles being broadcast out from our sun as well as from other stars, reaches our atmosphere and creates and animates life continually. In the ancient Indian studies of energy this is called 'Prana.' The other aspect of this light is that it carries the vibration of what we call 'Love.' It is pure in it's essence and loves all things. When beings encounter the vibrations from the waves of Prana, they can experience Love, Joy, Inspiration and Creativity. It is a personal expression of these higher frequency vibrations. We all use this pure energy here in our world creatively to express a wide variety of different forms from this Love. Before anyone in the moment realizes that they are using this Universal base creative and loving energy, they are already immediately channelling it through their life moments as a wide array of creative and loving variations.

Everything from the highest to the lowest manifestations and situations are created from this Love beyond our understanding. Some masters have expressed that everything is literally an expression of Love even at the lowest levels. If we were to understand at any given time that the quantum building block particles that animate life are not only purely creative but also purely loving, it could shift humanity into a mass awakening to a greater expression of living our existence.

In previous cycles of unity, Earth humans have understood and lived by this principle. Many tribes of the Earth are still in touch with this ancient knowledge and wisdom. The Atlanteans and Lemurians were consciously aware of this principle and living life by this standard. As separation started to occur the lower levels of love such as war and greed started to consume humanity while money, lust and power took over much of the group ego of humanity during the great age of duality and separation which we are just working our way out of.

Everyone wants to know when that is going to happen. It's been happening as the world continues to change through multiple events. The closer we get, the more conflict is arising within the group herd and individually as the duality bubble gets ready to pop or dissolve. Jesus had a made a remark concerning this future in the Bible: "I come like a thief in the night." We're all looking for that major, major shift that says to us 'Now...now we've gotten somewhere!"(ego?) Secret note to Self: *"It's always available and happening within and all around us."*

We are presently on the cusp of shifting back into the next cycle of Unity as it gets darkest before the dawn. If we can remember this dynamic of the Light and Love energy present in All things constantly without end, it can help those of us who are ready, willing and able to make the jump back to Unity and support everyone shifting into a more loving and creative expression of life. This is Heaven on Earth. Besides all of the colorful circus-like variations we experience here on the Earth plane, there exists other alternate or parallel realities where different variations of this reality are playing out. Some better, some worse and some just different.

What's more, these mirror realities can influence or overlay our present intended reality causing it to shift whether individually, by group consciousness or on a world scale. And if that weren't enough, we have a whole spectrum of different dimensions also influencing our world. Just another one of many reasons to be kind, loving, forgiving and grateful. In these other dimensions the light becomes distorted or mutates.

From the higher vibrations in the multi-layered frequency spectrums down the lower there are transmutations of dimensions and realities that occur outside of our known agreed upon reality. So dark beings or entities are some of these transfigured energy forms. And remember God loves everything and everyone. Those beings can receive the Great Light, so they continue to scurry around in the shadows creating lessons for people. At that level, things get really interesting. This is why we have the Arc Angels, to work as guides, referees, coaches, security guards, police, teachers and healers.

We are presently on the cusp of shifting back into the next cycle of Unity as it gets darkest before the dawn.

Meditation Exercise 4

THE SACRED CIRCLE

Meditation for receiving support, messages and healing from the Masters and Angels. This meditation can be augmented as needed. Learn the process first.

Always remember to never do this work while driving a vehicle or using any type of machinery. You must always be stationary in a quiet meditation. Read the script and then close your eyes and visualize the journey.

Relaxing yourself in a safe and quiet space, start by taking some long slow deep breaths. Let all the cares and concerns of the world melt away. This is a safe space and this is your sacred place. As you continue to relax and breathe, scanning now your whole body and becoming aware of each part of you now relaxing. Continue this for a few minutes until you are fully relaxed.

As you find yourself on a soft sandy path walking along, there are very tall trees on each side of you. You are now coming to an opening in the trees where the path opens up to a view of the ocean in front of you.

The temperature is perfect. As you step gently forward overlooking the ocean you are standing on a beautiful scenic overlook which slopes gently down to the beach.

As you sit gently down cross legged on the ground you now call up from the ground your Sacred Holy Fire. This is your True Essence and your Life Force Energy.

As you focus on your crown, this is your connection to all of the higher beings and dimensions. Breathe the Holy Fire in front of you to connect it with your violet Crown center.

As you focus on your 3rd Eye brow center, this is your connection to projecting forward onto the path of life in front of you. Breathe the Holy Fire in front of you into your indigo royal blue 3rd Eye.

As you now focus on your throat center, this is your connection for speaking through sound your truth and voicing your creations out into the world you. Breathe the Holy Fire in front of you into your sky blue Throat center.

As you focus now on your Heart, this is your connection for feeling love and projecting it out to everyone and everything around you. Breathe the Holy Fire in front of you into your emerald green Heart center

As you focus on your Solar Plexus, this is your center for personal power and moving forward through action and awareness. Breathe the Holy Fire in front of you into your golden yellow Solar Plexus.

As you focus on your sacral center (navel area) this is your center for projecting creativity out in your life and the world around you. Breathe the Holy Fire in front of you into your bright orange Sacral center.

As you focus on your tail bone or Base center, this is your foundation, stability and abundance that you project and connect with out into the world around you. Breathe the Holy Fire in front of you into your ruby red Base center.

Let the Sacred Holy Fire of your True Essence in front of you ignite the Holy Fire in each of these energy centers, strengthening, clearing, healing and balancing you. Breathe the Holy Fire in front of you into all of your chakra centers now together as they all resonate with this energy. Restabilzing, Energizing, Healing and Clearing all Chakras.

As you now call around You the Ancient Masters, Teachers and Angels to sit and hold council with you. To bring their messages and their energy vibrations to you.

We ask the Presence of Creator God, the Universe and Mother Earth to join the circle first. Bring your energy and special messages to the circle as we listen.

(pause for a moment and listen)

We now ask the presence of Jesus to join us. As he comes he brings the great Unconditional Love that he showed to all people. As we listen to the truths of his messages.

(pause for a moment and listen)

We now ask Mary Magdalene to come to the circle. Bringing with her the Divine Feminine Christ vibration messages as we listen and receive.

(pause for a moment and listen)

We now ask Mother Mary to come to the circle bringing her Blue Light and the Grace and Holy Light Essence. We listen to Mary as we receive the Grace and Miracles coming forward to us.

(pause for a moment and listen)

We now ask Babaji to join us, bringing the spiritual truths of yogic mastery, Great Truth and wisdom as we receive his messages and vibrations.

(pause for a moment and listen)

We now call Sai Baba as he appears and brings his manifesting magic of creating gifts and messages of Love, Wisdom and Inspiration.

(pause for a moment and listen)

We now call Paramahansa Yogananda as he brings his gifts of Love and messages to us for uplifting our life.

(pause for a moment and listen)

We now call St. Germain as he brings the Sacred Fire of the Violet Flame to clear our energy and move through us with the Holy I Am Presence.

(pause for a moment and listen)

We now call Serapis Bey as he brings his knowledge and wisdom of messages and blessings from Ancient Egypt.

(pause for a moment and listen)

We now call on Kuthumi the Holy Saint from India to bring us his wisdom and love.

(pause for a moment and listen)

We now call on El Morya the master from Ancient Tibet to bring us his Tibetan mystical healing and messages.

(pause for a moment and listen)

We now ask Quan Yin to Join the circle bringing her compassion and rose quartz love to lift you up. Listen to the Heart of Quan Yin.

(pause for a moment and listen)

We now ask Master Jwahl Kuhl from Tibet to join the circle bring his ancient mastery and wisdom from the mystical traditions.

(pause for a moment and listen)

We now ask Merlin the great ancient magician to bring his magic and messages to lift up and transform our reality.

(pause for a moment and listen)

Let's now ask any Native American guides or Animal Familiars guides that are helpful to join the circle.

And now let's ask your personal guides and any ancestors or family members that want to join the circle. We are thankful for your presence and your messages.

(pause for a moment and listen)

As we sit quietly lets receive healing and messages from the Great Ones. If we have a question we can ask it to them. As they all sit with us around our Holy Fire.

(Take time to receive their vibrations, messages and answers to your questions)

Appearing from above is a bright pillar of light. As you look up into it, the light becomes even brighter, there is a soft quality, a warmth. As the pillar of light comes down like a spotlight to where you are, the Keepers of the Akashic Records are granting you access to your Book of

Life. As your Soul Records are opened from above.

We call down your original Soul Codes for healing and regeneration of your original DNA.

We call down for you the original DNA blueprint to over write your existing DNA lifting your energy bodies and your physical vibration.

We call down for you now the cellular DNA into specific areas of your body that need attention. See the light flooding into those body areas as you focus on them one at a time. Healing, Balancing and Energizing and Activating You.

We now ask to download the Unconditional Love from your Soul in Heaven. The Unconditional Love now comes flooding in and clears, and heals any and all previous lower vibrational thought and emotional forms that had been stuck in your body. As these patterns leave you, they are washed out to the sea. See them leaving.

Your vibrational light is now shining and resonating with Harmony, Good Health, Success, Love, Creativity and Strength. With all these gifts that have been given to you, allow them now to integrate and rest assured in Body, Mind & Spirit that your world and your life is transforming for the better.

Coming back now, your nervous system, your heart, your mind and subconscious are all aligned with all that is good. Remember this, walk this and live this. Enjoy this in every moment as you move forward with certainty from within, grace and balance.

As the light starts to move back into the heavens and

you are there with the Masters all gathered to acknowledge your success. It is time to bid them farewell for the time being as your Holy Fire begins to fade, but as it does it moves into you.

It will go forth with you and you will see it again when you return to this Sacred Place. As you get up now turning you walk back into the trees along the path and come back home as you walk back into your present self. We thank the Ascended Masters and the Keepers of the Akashic Records for this Sacred Circle Healing and Activation of our Divine DNA. Om Peace Amen.

Chapter 8

THE REMEMBRANCE OF HOME

In your journey through life you have silent questions within you operating from behind the throws of everyday life. As we project our lense of consciousness out into the external world we are silently asking for these great unknown questions to be answered. These silent questions can be a driving mechanism in our interplay with the world. We search for love and fulfillment. We search for inner peace, comfort and safety. We search for a greater sense of creativity and abundance in our lives. And above all, through all of these things, we search for a sense of Self Knowing or Self Realization whether we are aware of it or not.

There is a part of us that remembers what it feels like to be whole and complete. No matter how challenging our lives have become we know there is more. Many of us move through life with the great waves or tides of creation ebbing and flowing on the physical plane. Jobs, community, family, aspirations, goals, dreams, projects, healing and so much more. The ways are never ending and ever beginning here on the Earth. As a way of navigating with minimal struggle or suffering we ride these waves through learning to balance. Letting go and accepting at the same time brings us a creative way to surf the tides and continue to live a somewhat balanced existence. Some of us have mastered riding these waves of creation and have come to live a more prosperous, happy and fulfilling existence.

While others are at the other end of the spectrum are still learning to swim and seem to have more of a challenge in resisting or fighting against these great waves. (See the illustration of waves) While a great number of us have learned to move in rhythm with these waves in one way or another. As we continue through life we generate strength through our flexibility with these waves. Like the great trees our strength to navigate life can often be in our flexibility. They are rooted and grounded though they bend and dance with the wind and the weather. Waiting for their day in the sun.

When the wave patterns in our lives are more pronounced and the waves are taller and deeper, the patterns of life can be more dramatic. As we can become more present to this dynamic we can learn that there are times of action and times of rest (which can also be an action). Our possibilities for harmony and balance increase greatly when we can learn to even consciously decrease the waveforms into a more shallow flow. This is when the roller coaster effects in life start to calm down and with cause we affect the size of the waves in life, as they are still ebbing and flowing yet more consciously and harmoniously moving and dancing with our world.

Some ancient monks developed the practice of Tai Chi around this concept. They were ebbing and flowing, dancing in harmony with nature whether for protection against intruders or for meditation and joy. Their poses were taken from nature, thus watching the movements of the animals and creating martial arts poses based on harmony and balance with all life, the Universe and nature.

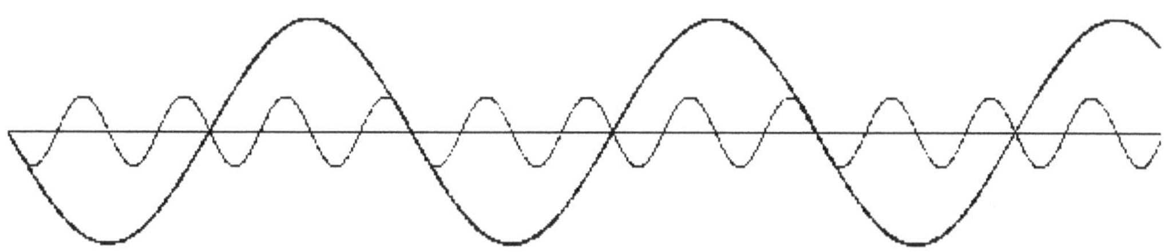

Here are some Examples of Harmony and Energy Balance:

1. Knowing when to not overwork ourselves or take on more than you can handle.

2. Taking better care of our health with better diet, exercise, hydration, supplements, and getting enough rest.

3. Cutting back or stopping excessive alcohol, caffeine or drug consumption. Or any other life altering habits.

4. Becoming aware of our emotions and practicing forgiveness, gratitude and humility.

5. Spending time in prayer, meditation and

visualization.

6. Spending time reflecting on our progress.

7. Learning to let go and let God. This is a key factor to the ebbing and flowing in harmony. Listening to yourself, the world around you and to the God within you.

You would find if you were to study the martial arts and, there are many practices. You would witness the aspects and elements of the practices that were lived and taught by those who became great effective martial artists.

Tai Chi and Chi Gong are 2 very worthwhile and easy internal martial arts that can help you develop your relationship with energy, both inside and around you. I often recommend these 2 forms to clients who want more energy flow in life and to develop themselves more.

We are then starting to answer some of these unasked and unanswered silent questions within us. This is also when reflection gives us an understanding that we are making progress, no matter how much the ego tells us that we are not. We start to develop an inner knowing that continues to grow. A faith within ourselves. It is a special feeling when we are able to navigate through a situation much more easily than we ever did in our past.

It is important to note that if we do start to experience growth, movement or positive changes in our world and we come to a place of inactivity or sluggish progress not to become alarmed. We don't want to impact a slow time with feelings of failure or fears of backsliding. Always keep your head up and your eye on the ball. All things in moderation. This is how the turtle won the race against the rabbit. Slow steady progress and growth (with some eminent victories along the way of course) builds character and wisdom. Ride the waves.

The Great Masters through time found their own peace, love and balance as they were able to hold it within them and around them until it became part of their dominant character projection on the physical plane. It was then that the world, all beings and the Universe answered to this divinity. They projected what would manifest through the Great Mirror of Reality. Often times creating things or healing in such a way that seemed miraculous.

Some masters travel by thought or bi-locate. Disappear and then reappear at another location. Driving out dark spirits or changing the weather. All because they are in complete harmony with nature and the Universe. This is truly how the miraculous is achieved. By being in complete unwavering communion with nature, the world and the Universe. This is when they become both the wave and the straight line, a constant. Complete unfettered harmony. Astrological anomalies or processions have no affect on there lives because they are one with them and operate just as interdependently with the Universe as the stars and planets themselves do.

You are that master. We have all come here to master

our own existence. Whether it's a normal everyday life or we augment it with some forms of spiritual or mystical studies and training. Before we have come to the Earth while still in the Great Halls of the Akashic Records, we chose the things before our arrival that we would like to study, experience, heal, create and complete while we are here. Once here as an infant and then a child we had forgotten our heavenly choices before coming. As we learned to navigate through our early lives, the situations we moved through with family and development were all part of this plan.

We tend to generate more thoughts and emotions in this circus life reality than we know what to do with. Sometimes causing different unintended karmic patterns to arise, though nothing is waisted, and everything is used for our growth. Those Souls who have already developed over a longer series of incarnations can usually find their way a little more quickly into a personal knowing than those younger Souls who are still evolving.

This is the Great Dance. Billions of evolving Souls at different levels and intervals, all interacting with each other and some of the Soul mates they recognize along the way through family, relationship partners or even conflicts that arise. All, and if only, to generate a response. And the response that the Soul, the Guides and Heaven are looking for is the greatest response of Love or Healing, Harmony and Completion. We do have free will down here as to how we respond to our world and this mirrors to us and through our Book of

Life in the Akashic Records our own growth, progress and completion to certain lifetime dynamics.

So work on yourself, as you are the Master in training. You have the same mind, heart, DNA, chakra system and Soul setup that all the great masters have had. Learn to navigate the Earth vessel and body temple wisely. Be all that you can be in each moment by becoming more present in each moment. The past holds many regrets and the future holds many fears for most of us. Be present right here and right now and that helps to shut down the dynamic of the ego's hijacking of the subconscious mind. As it tends towards manifesting or playing out endless new versions of old scenarios based on negativity, regrets and fears.

Now if we are more present we can naturally call forth from our subconscious and inherently higher aspects of the ego that can work in harmony to bring about new versions of positive dynamics we may have experienced in the past such as creation of some masterpiece or a major win of some sort. The positive always out weighs the negative in this way. And remember your guides and the angels are always shining Great Light and Love to you here so you are not alone in this. It has been said many times, 'If we could only see who walks beside us we would never question ourselves ever again.'

We are all one, in a way, through Universal consciousness and yet we have a separateness or individuality within it. This is the Soul and also this is the human body in a physical world. So both dynamics are true. We tend to shift in one way or the other from time to time, while both dynamics are actually constant and equally true.

The Remembrance of Home

We all have within us a silent memory of home, heaven or the other side. Through our memory being wiped that happens when we return to the Earth through birth, we are mentally disconnected from where we just were, in the Akashic Records as part of Heaven. Though we carry with us deep within our subconscious minds and our Soul fields the Inner Knowing that we are not from around here. We come to the Earth plane by our own choice to learn, evolve and grow.

The dualities of thought and emotion start to work within the confines of worldly dynamics of duality and the show begins. These dualities include the influences of the group consciousness of heart and mind of the others incarnate in human form as well as our own unique perceptions of our reality through our heart and mind. This starts the interactions with the World Mirror and the Universal Energy Mirror as our reality. This mirroring creates 'cause and effect' or varying levels of karmas.

Again, we are influenced by Self and by the Group Self. Group Self influences can include what the world is like at the time we are born or what is expected of us within a certain race, country, region or state. It could be influenced also by nature and weather, etc. Personal self influences are created by how we respond and interact with these group self influences as a whole myriad of personal interactions with others, what we choose to create through each moment with our studies, tests and free will creative choices.

Whether or not we ever find out about the mirroring effect is inconsequential to leading our lives. All the better for a higher rate of learning and expansion if we do. There are a plethora of idealisms, and instructional sayings around life's values and purposes that are in alignment with this mirroring effect that most of the world grows and learns to live by:

"What you sow is what you shall reap"

"The journey of a 1,000 miles begins with one step

"Know thyself" ,

"If you can't stand the heat, get out of the kitchen" ,

"This too shall pass."

Here on Earth we adopt these sayings of great teachers and philosophers as part of our everyday guidance. This helps us to have a focus as we are experiencing Earth life.

In current times we have more information available to us about the art of manifestation, living a successful life and mental and emotional healing than ever before. This can really help with navigating earthly lessons that we set up for ourselves prior to coming down.

As we learn about the Akashic Records and start developing our abilities to look into our own Book of Life we do this through our Soul. Our Soul is the part of us that came from home or Heaven sent down to learn, experience and grow. When we connect with the Akashic Records just by way of reading books, learning about it and doing guided journeys into it, we wake up our Soul field connection within our conscious mind/ heart and body awareness.

As your Soul field becomes naturally present or activated with the physical body through your mind, heart and energy fields you will start to experience a unique level of fullness or connection that is subtle and yet pronounced. Your True Self is now coming online within the Earth reality. It has always been with you silently observing, experiencing and learning through reactions, responses and situations.

This added kinship with your Soul connection gives you an added view of personal self growth and study. Now you start becoming present to more of your life than previously, because you are now aware that you are participating in a reality creation game of 'call and response.' This can help us to avoid knee-jerk reactions to life situations with a deeper understanding. Once a different response is given the rules of the game can change. The world, situation and reality mirror reacts or responds to your reaction or response creating a new outcome. This changes your karmic patterns.

All the while you are starting to understand and become more aligned with your Higher Self. You are not hacking into heaven or the Akashic Records, you are simply aligning with a silent connection that's always been available to you. This can be a game changer for you.

Very few people in comparison to the general population ever make this connection though they are participating just the same. Their Book of Life has within it all of the life experiences that they ever signed up for and their creations and creative responses in life on Earth while they are here being documented continually in their Book of Life. Knowledge of this has been lost over time. So here we are helping you to remember once again.

If a person does nothing in this regard they will be fine, they will report back to Heaven or the 'real world' after their transition and do their Life Review. They will decide whether there is a sense of completion in study or if they want to come back again. In some cases coming back can be strongly suggested due to some karmas. If a person can study the Akashic while on the Earth plane and look into the Self Knowing aspects that it has to offer, they will be ahead of schedule in their learning.

Some of us find the Records through dreams or near death experiences (NDE's) where they are taken up and shown something, while out of body during a traumatic accident or illness that creates an opening for the Soul and the conscious mind to step through to the other side for some interesting learning, healing or needed karmic direction. Many people who have a direct experience like this come back with a different frame of reference for life. The goal is healing and expansion.

People often come back from the other side seeing God or Love in every person more than ever before. Some return with expanded psychic awareness or able to see auras. Some people report even being able to see the others on the spiritual planes present around us all.

In my work with the Akashic Records I've explained the dynamic of continual connection to it through study, self study and practiced guided meditations as a subtle prolonged version of an NDE. I like to refer more to it as a Near Life Experience. I am different and I am changed as a result of doing this work for the better. Some may say. 'Well you are going to change in life anyway over the course of time.' And there are other ways of doing the healing work. All true, but when you know, you know. When you have gradually awakened as a result of continued visits to the Temple of Souls, you just know it from within. No one can take that away from you. There is nothing quite like doing the Akashic work to move the energy on a Soul level when nothing else has seemed to work.

The more you look into your Book of Life through your Soul Connection and consciousness, the more you start to subtly change and transform your life. Your psychic intuitive abilities will naturally start functioning. So many have asked "How do I open my 3rd eye?" Well there are many ways. But do you know what you are connecting to really? Having a target location for practice is best and safest both karmically and energetically. Looking into the Akashic Records is a completely protected safe space to work from.

When I am working with others in the Records, or I am teaching workshops or even writing this book, I feel a sense of being connected and feeling complete. While I am here doing this work my Higher Self is actually working in the Akashic Records as a Keeper of the Records. We all have different creative lives and partners on the other side in Heaven. This is my Remembrance of Home.

Soul vibration and the vibration of Love, Healing and Rejuvenation. So this Remembrance of Home can be a 'go to' for you. A place within yourself of a greater purpose, Your Soul's Purpose. I have worked with people in session and classes who struggled with holding on to family emotional hardships their whole lives. By doing the Soul Records Access Work and healing exercises, they found their pathway into understanding and completion with many of those hardships, if not all of the issues. They have reported healing and that the issues no longer affected them. Physical healing has also taken place often.

The Great Halls of the Akashic Records, The immense Love and Healing, the Angels, the Everlasting Life, The Ascended Masters and God. It's all real. This is my Remembrance of Home.

Chapter 9

SEPARATION & UNITY

Separation -
- The action or state of moving or being moved apart:
- The division of something into constituent or distinct elements:
- The process of sorting and then extracting or removing a specified substance for use or rejection.
- The process of distinguishing between two or more things:

Unity -
- The state of being united or joined as a whole
- The state of forming a complete and pleasing whole, especially in an artistic context: the repeated phrase gives the piece of art unity and cohesion.
- A thing forming a complex whole.

The two overlaying truths or laws that govern our perception internally and externally are that of unity and separation. These dynamics govern our conscious awareness individually as well as collectively. It can move out from the Self in perceptive delineations of family, friends or foes, community, region, countryside, state, nation, continent, race, civilization or planet. For us, we all have an agreed upon recognition of certain outer groupings. We live our lives by recognizing the differences and similarities between these groupings. From the most mundane levels of knowing that California exists next to Arizona and Nevada. Or that Spain is next to France. Germany is next to Austria and so on. We agree on, or have a knowing that a certain grouping of communities exists within a state or country. Even so we have quite a lot of regions that are unexplored on our planet alone, so less is agreed upon in these areas, except that the said area or region exists.

As we move through our life moments at the most basic level we have our first frames of reference which usually pivot between our inner world (what's going on inside us) and our outer world (what's going on in our reality around us). These are our most basic instinctual points of reference. Our ego and the lower mind through the use of our senses and nervous system tend to operate here.

As we continue forward in our spiritual studies and our personal healing work we have yet another perspective as we start tending into the Higher Mind and the Heart as well as becoming more aware of the energy fields while we tend upwards and expand to connect with God, the world of Spirit and the Angels and Guides. It is here the we can connect with the Akashic Records.

We also then have our sensory perspective of ourself in relation to another person or groups of people. Our differences and similarities. This is usually unconscious yet at times a conscious perception, and is always running in the background and... at times in the foreground of our lives.

As we are attempting to make sense of it all through our growth and awakening process, we can find ourselves shifting back and forth between our sensory driven Ego filtering lense of life and the wide angle lense of connecting self to a higher spiritual presence. At times we then begin experiencing all of this, the separation and the unity perspectives simultaneously. Even more confusing (hopefully not always) is the shifting back and forth between these perspectives while we are still in the same moments of time, the same surroundings, with the same people and so forth.

Many elements of our outer world do not change and continue to run their course as we are experiencing changes or shifts in perception back and forth from one frame of reference to another, or experience multiple frames of reference simultaneously from our inner perspective projected out.

Several things are happening when this occurs. Our energy fields are expanding and allowing more of the Universal consciousness to run through us in Mind, Heart, Body & Soul. Our nervous system which also projects our emotions and subconscious directives becomes more sensitive and interactive within and without. Our endocrine systems and our chakra systems may start to exhibit special gifts allowing us to do certain new things which were previously not available. So you have a greater knowing and perspective of how you are connected and everything starts to work both individually and together.

As we work at maintaining our steady flow of life events these shifts, whether noticed or not, can account for a myriad of energy and reality fluctuations that we are both directly and indirectly experiencing and either we are actively aware of or silently unaware of. These are stops and starts in the reality flow. Projections and emanations that create reality. Quantum waves of high frequency slowing down in a certain spot to create a unique event, pattern or manifestation. This can even result in a shifting of the realities altogether or the slowing or quickening of manifestation patterns along the group and/or personal event horizon.

So as we're just starting to figure all this out by being present to it all and continuing, we start to develop an inner knowing that we have a working model that is simpatico for our understanding towards a successful life. The ego is now shifting gears to work more in unison with the Higher Mind, God, Higher Self and the external reality..."Ok, I have a working reference point." Just when we think we have it all figured out, someone interjects "Yes, but It's All One!...It's All Love!....We are All Connected!" smiling brightly.

"Okay, Hmmm, wait a minute," the ego says "You mean I'm connected to adversaries, opponents, past partners, fair weather friends, an ex boss, or political leaders of the 'other party'?? "Why Yes!" says the ecstatic Higher Mind & Heart shimmering with glee. Meanwhile, the Unconscious Primal Self does ***Not*** jive with this at all. Though the Higher Mind and the Heart know that it is in fact All True. As well as the Soul which is the True Self. The Soul is connected to a Soul group, and then that group to the other groups in a great pyramid scheme matrix that goes all the way back to God. "Wait....What???" you sputter, as you are trying to micro process the macro immensity of the All. So we are all intimately and ultimately One.

And yet, the Soul is also a unique individuation or fingerprint of the Universal God Source Light of Creation. It contains, You, The True Self which contains all of the knowledge wisdom and energy that you could ever hope to find. You are truly Divine. And your individual Soul is interdependently connected to all other Souls.

Also your interconnected individuated Soul of higher existence is paired with your lower denser physical body, Ego and Unconscious Primal Self! *(we're capitalizing ego to make it feel special for a change)* And...they are both using the same senses, the same brain, the same nervous system and the same chakras. "You are being hijacked, you are being rented out!"... the Ego says.

You are individual and You are One with everything and both dynamics are true and happening at the same time. This is the great synchronistic paradox of reality. Unity and Separation.

Pertaining to the Akashic Records this happens on a whole different level. As Souls we all come down to the Earth plane to experience life here in this dimension of duality with all of it's many joys, woes and everything in between. The Earth School really is a quite diverse playing field. After we have reviewed all of our previous life's triumphs and missed opportunities we pick and choose the experiences that we think and feel will give us the most growth on our next Earth life incarnation visit. This can be a wide array of events. We choose where we will be born and to whom. Often we choose parents or family members, reincarnating Soul groups and Soul mates. This can often be in the same family so that we can have the most Love and creativity with, the most learning and/or the most healing with from past life events. We choose what we will accomplish and achieve. We choose our own stumbling blocks as a way of learning, testing or even healing past life karmas.

Sometimes we choose a certain illness for the same reasons. Birth, partners and how we will transition back to Heaven are the key Akashic elements, and then we choose every other major experience we wish to have. These are in our Akashic Book of Lifetimes stored in the Akashic Records. These key element life events come down with our Soul as we become enjoined with a physical birth. This is where the question has arose: 'Do we actually have free will or not?' Yes, we have free will. We have free will to have coffee or tea. We have free will to wear the clothes we want to wear, to have holiday activities and play a sport. To play music, write or paint, or to travel about freely and so many other choices about relationships and career.

Everything is free will up to point that any major decision would change our life path in such a way as to create conflict with our Soul's Purpose. Which is our

Soul and it's connection to our Akashic Book of Life comes and guides us directly into a sequence of events that plays out the major choices we have made before we came down from heaven. If we make choices out of 'free will' that are out of harmony with that scheduling process, conflicts may arise to get us back on course. Now there are sometimes unforeseen events that play out on the Earth plane with group consciousness and this also can play a part in our movie. This can result in life path detours.

Here is another important factor. Some of us start to wake up on the dream scape of Earth life. We are learning about spirituality and emotional healing. Karma cleansing and better health choices. So this creates a positive spin on a major event we may have set up for ourselves to play out and it is now being changed or rewritten for us as we are levelling up in a new way. This is not breaking any rules. They *(our friends upstairs)* want us to progress and do better. If we bypass any pain and suffering by taking an augmented journey through healing all the better. We are able to arrive ahead of schedule or keep our schedule more clear as we intend healing and inner knowing for ourselves. In some instances a similar event could present itself but with much less magnitude. You are not hacking the system you are simply levelling up more gracefully in the great scheme of things.

If you were to look into the Records and see a coming unfavorable or difficult event we can go to work through prayer, spiritual study, emotional healing and karma clearing and in many cases change or even clear that coming event. Because it will either no longer be needed to get a certain set of responses to generate learning for us, or it can be greatly lessened.

When you find your way to the Akashic Records and you start studying it and learning about the dynamics of your Soul and it's Purpose for your life through looking into your Book of Life you are now becoming more fully aligned and connected with a broader scope of the true reality and the way things work. This is when we start to change here on Earth and become more imbued with our Soul's energy. 'On Earth as it is in Heaven.' This alignment creates more Unity between you and your Higher Self.

This creates less of a need for a separation of coming to Earth and living your life's lessons while being mostly unaware, and then reporting back at the end of your Earth life to enjoy Unity again. This new shift's intervention is huge! This lessens the need for more lessons and opens life up for different patterns and manifestations, as well as, creates an integration between the two worlds instead of the old need for a complete separation.

And when one awakens, and then another one, and then another, in this way we are not only changing our lives, we are bringing more awareness for more of the Souls to awaken on the Earth plane in a safe and beneficial way for everyone. More lights on the Tree of Life or the Soul Tapestry of Humanity will be glistening.

Other perspectives have included ideas such as

'Everything is as it should be' and 'You simply do not need to do anything.'*(generally these statements can be based on Zen Bhuddist approach to reality)*These statements are coming from the idea of releasing conflicts through non-judgement, which is a unique approach. In context to what we are talking about with energy studies, practice and expansion: While those general statements have truth, it can be a condoning message for those who are not immersed in Self or Soul studies, but find out about the Akashic principles and start to wonder more about the consequences of their own lives while seeking a general level of comfort and approval. It's also a good frame of reference for dabblers or the novice who wants reassurance. Those statements fall short of taking more responsibility to learn and put into practice the nature of who you truly are and who you came here to be.

If I may be so bold, I think we owe it to ourselves to understand deeper levels of reality and interact with our world through the best understanding of it that we can embody. It's not practical to think that everyone will 'get it' on this level. Although it is always available for anyone and everyone. And as many of the sacred texts have said 'it comes in the twinkling of an eye.' My passion is to help the Souls that are ready to find their way to this knowledge and help create wisdom as their lives play out for the better and the greater good. As you drink in more knowledge and practice, it will level you up to a new normal, and that can be transformative. That's what we are here looking for.

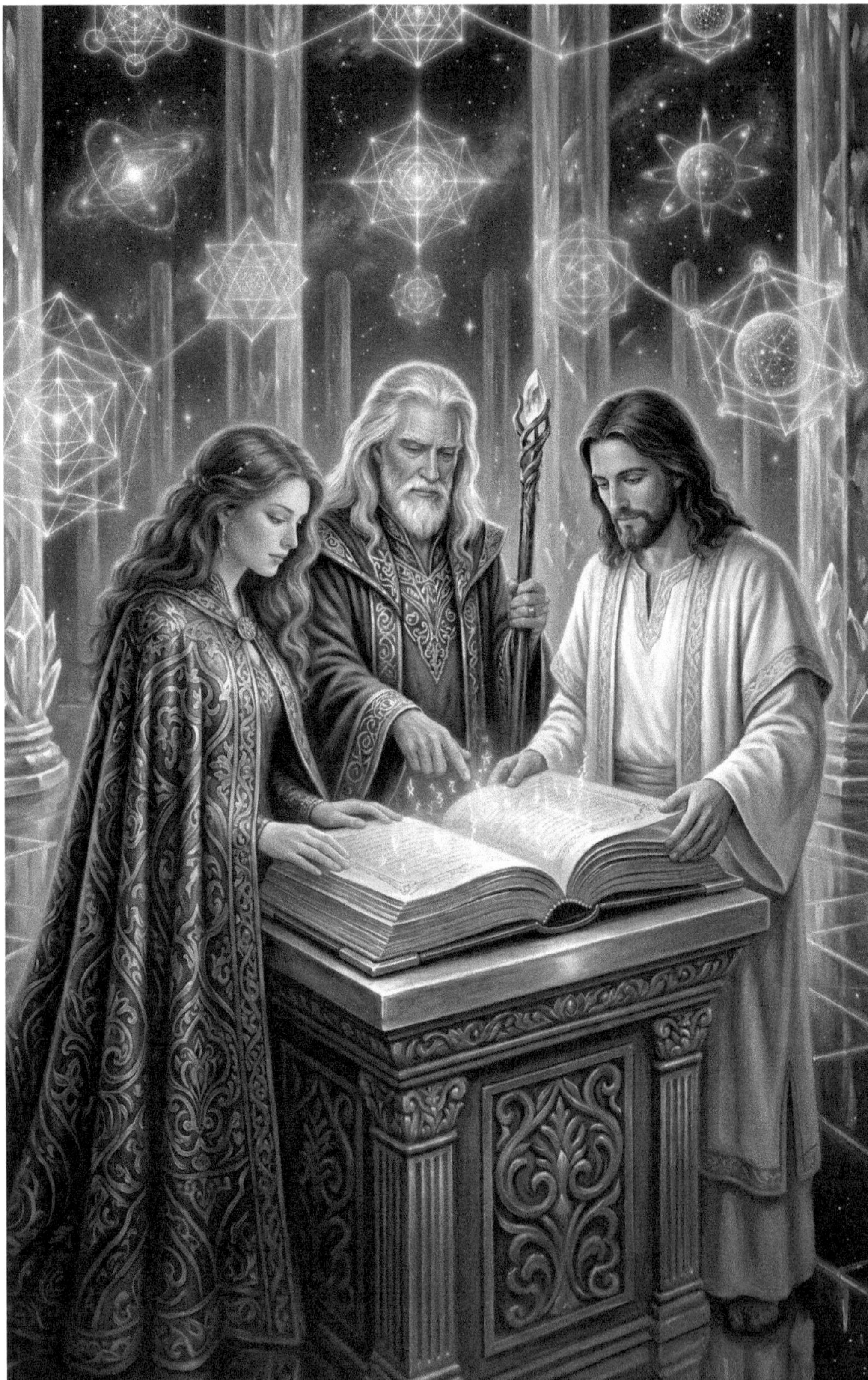

Meditation Exercise 5

Meeting your Guides In the Akashic Records

Guided Journey Meditation - Always remember to never do this work while driving a vehicle or using any type of machinery. You must always be stationary in a quiet meditation. Read the script and then close your eyes and visualize the journey.

As you sit or lay down getting comfortable, really let yourself start to relax more and more. Breathing in and out very long, slow and evenly. Just let all the cares and the concerns of the day fade away. Relax every part of your body from the top of you head to the tip of your toes. Become present with each part of your body as you relax that area letting it gently release. As you continue to relax your heart rate slows down ever so gently... and now relax your mind with gentle peace. Just allow it to really slow down knowing that all is well and as it should be.

Relax now your facial muscles, your forehead and your jaw. Allow your mouth, lips and tongue to relax. Allow your eyes, eyelids and your eye muscles to release and relax ever so gently. Relaxing your ears and your hearing. As you quiet and relax down even deeper taking the time for your body, mind and spirit, visualize a beautiful golden sphere of energy around your head and chest. A large sphere that extends up over your head. This is your Soul energy. As you see it, start to breathe this energy down into the top of your head. Breathe the golden light of your Soul into your face. As you breathe the golden light in now coming down into your chest and heart. As you continue breathing the golden light of your Soul down around your body and through your body until it fills you up completely. Fully coming into your body lifting you up gently and supporting and soothing you in every way.

Focus your mind down to your base chakra (ruby red) scanning it with your consciousness.
Continue to scan up into the sacral chakra (orange).
Continue to scan up into the solar plexus chakra (yellow).
Continue to scan up into the heart chakra (emerald green).
Continue to scan up into the throat chakra (sky blue).
Continue to scan up into the 3rd eye chakra (indigo blue).
Continue to scan up into the crown chakra (violet)

Now without scanning back down through the chakras, move your awareness back down to the base and scan a little more quickly up through all of the chakras. Once you reach the crown, circle back to the base and repeat the upward motion scanning. Do this a few times until you feel the energy opening and moving.

Find your self now in a beautiful outdoor setting. There is very lush green grass and large trees. The blue sky overhead is full of white puffy clouds. There is a gentle breeze. You can hear a nearby stream gently trickling. There are beautiful various flowers of all colors around you. You can feel the gentle breeze as you see the shade of the trees dance gently to the wind. Feel the plush grass under your feet and between your toes.

As you walk out into the sunshine you see a crystal staircase leading up into the sky. It is glistening with light and seems very much alive. You feel the stairs calling you. As you approach the staircase and you now touch the stairwell, you can feel the subtle vibration calling you to climb the stairs. So you decide to do it and you take the first step and then another and then another. Climbing higher and higher and feeling lighter and lighter. Brighter and brighter feeling lighter and lighter, climbing higher and higher. As you continue upwards the Earth falls away below you as now your feet start to glide up the stairway effortlessly.

As you look out you see the celestial heavens all around you. Climbing higher and higher as you see pure white clouds above you. As you float now up into the clouds the bright white takes your site away. But that's okay because your Soul knows the way.

Coming out of the clouds now, you see a beautiful white marble temple before you. The Great Hall of the Akashic Records. Your Soul is filled with joy to return home on this visit. As you now walk up the stairs to the front landing... 1- 2 - 3 - 4 - 5 - 6 - 7 .

As you stand now before a pair of great wooden doors and as you reach for the handle. Just as you do, the doors slowly and magically open to you. As you slowly walk in and look upon the Great Hall, you see very tall arched ceilings and stain glass windows lining each side of the Hall. The light of creation shines down through the stained glass and creates colored beams that shine across the Hall. You see an ancient book at the other end of the Hall with a light shining down on it from above. This is your Book of Life.

As you now take a few steps forward to walk across the Great Hall over to the Book, something catches your attention from the corner of your eye. Someone is there. As you turn to see them, it is one of your guides. Who is there with you? Take a few moments here to meet with them and have a conversation.

As you now start to proceed again towards the Book someone catches your attention out of the corner of your other eye. It is another one of your guides. Who is there with you? Take a few more moments here to meet with them and have a conversation.

As you now walk with these two guides over to your Book of Life they will help you as you look into your Book and to receive messages.

As you now ask the question "What is most helpful

for me to know from my Book of Life in the Akashic Records at this time?" As you soften your vision and relax your mind even deeper let the images appear in the Book or on the screen of your mind. You may see words, written messages, symbols, pictures or movies. Take a few moments here to receive the messages to the questions you may ask. "What is most helpful for me to know in this moment here in the Akashic Records from my Book of Life?"

Now as you look up from the book you notice someone else from your peripheral vision standing to your side. As you slowly turn your view, you now see who is there. Take a few moments to have a conversation with them, asking questions and receiving answers.

As the three guides are there with you now in front of your Book of Lifetimes, you may have a discussion with them to further open new information to you. Do you have any questions for them about what you are seeing in the Book of Life? Would you like their help and guidance in reading your Book? Take a few more moments to talk with them.

With new insights and a newfound connection with your guides here in the Akashic Records, while giving thanks you start walking with them back across the Great Hall trading stories and comments. As you now reach the front door and it again magically opens for you, two of the guides will remain for a while in the Akashic Library to do some study work. One of your guides will escort you back to the Earth plane as you walk out onto the great marble landing facing a spacious sea of white clouds.

As you step back down the 7 marble stairs that lead you to the stairwell. Take a moment here and turn to look at the white marble Temple of Records once more noticing all of it's detail and where you have just visited. As you step down the marble stairs 7 - 6 - 5 - 4 - 3 - 2 - 1 . You are now back onto the stairway as you glide with your guide back down through the clouds. Coming out of the clouds you see the Earth becoming larger and larger as you glide down the stairway. Coming back down to your continent, coming back down to your country. Coming back down to your city or town. And now coming back into your home and your body. Take a few minutes to reintegrate more fully back into your physical body. Wiggle your fingers and toes and move your body. If it's midday you may want to get up, stretch and maybe walk around a bit before you journal what you just experienced. If it's evening time and you are going to head into sleep maybe you want to voice the protocol "Dear guides please connect with me during my sleep and bring me healing and messages in my dreams."

We thank the Keepers of the Akashic Records for opening your Book of Life to You and for the meeting with your Guides. Om Peace Amen.

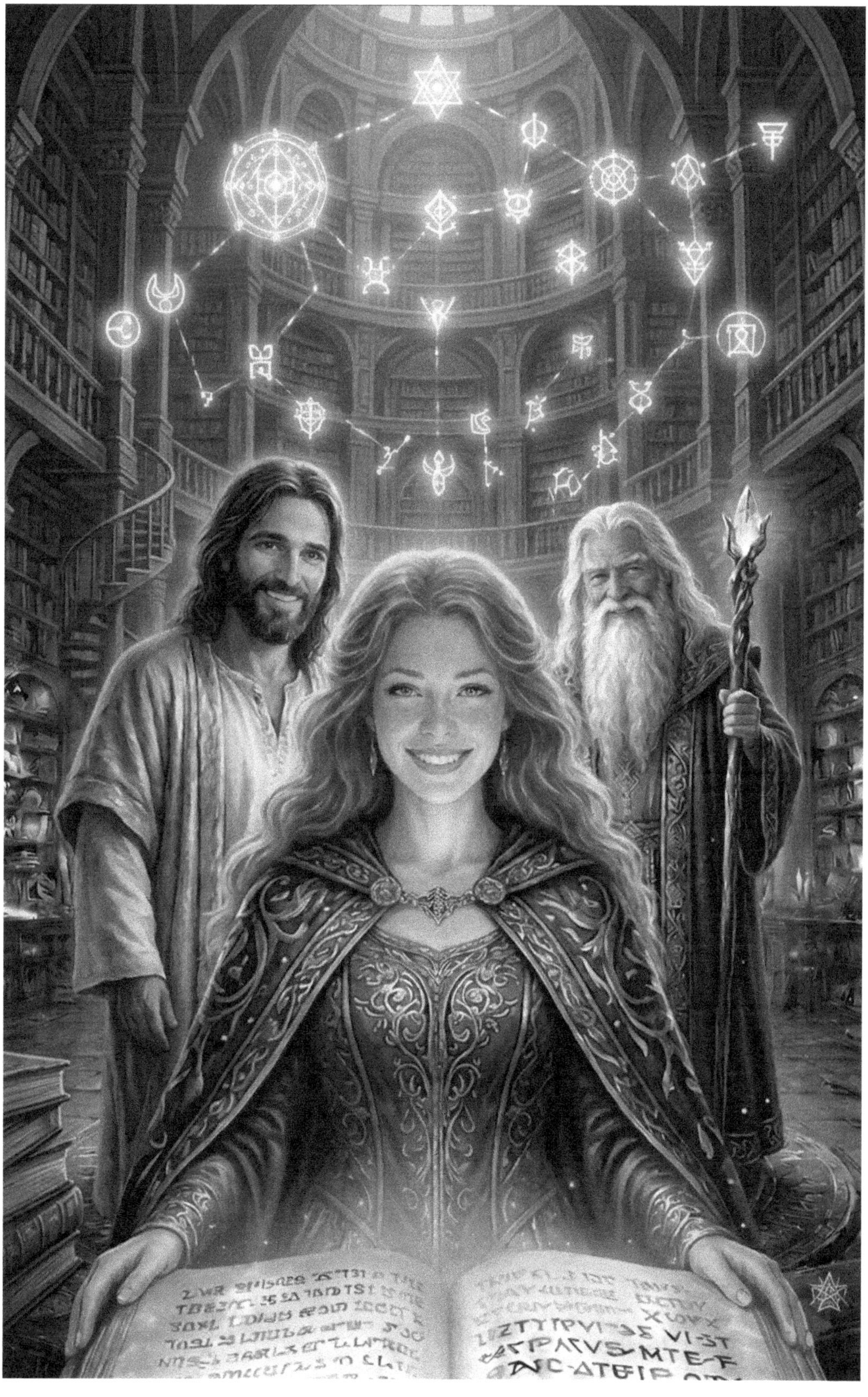

SURPRISE!

Chapter 10

SYNCHRONICITY & SERENDIPITY

Synchronicity - Describes meaningful coincidences where external events align with internal thoughts or feelings without a causal link, suggesting a deeper, interconnected reality between the psyche and the world, often felt as spiritual guidance or signs, though skeptics attribute it to confirmation bias or pattern-seeking in random chance. It's a "falling together in time" of inner and outer experiences, like thinking of a friend just as they call, or seeing repeated numbers or symbols that feel significant.

Serendipity - The occurrence and development of events by chance in a happy or beneficial way. Making fortunate discoveries by happy accident, finding valuable or pleasant things you weren't looking for, or experiencing beneficial events by chance. 'A fortunate stroke of serendipity.'

Through the concepts of Separation and Unity and the different levels, dynamics and delineations that we covered in the last chapter, we now look at Synchronicity & Serendipity as another interplay of our conscious awareness. When our inner and outer worlds sync up or align, things happen naturally for us. Though it can bring varied results based on what's happening in our inner world, let's focus on the good.

From the ancient teachings their are quotes that point to the instruction of synchronicity. There are so many sayings, proverbs and parables that exhibit the same quality quite readily. Here are a few:

'As above so below'
' As within so without'
'As a man thinketh in his heart, so is he'
'If anyone should say to this mountain, Be thou removed and be thou cast into the sea, and shall not doubt it in his heart, but believe these things he speaks are true, he shall have these things'
'To be on the path you must become the path'
'To him who hath it is given, and to him who hath not, it is taken away. '

All of these quotes talk about what we experience on the inside and what we experience on the outside (in our reality) as a result of our inner thinking, feeling and being. So they actually point to the alignment with manifestation of that which is desired...or undesired. So much, that it could be said the whole idea behind 'synchronicity' is that of the law of attraction, manifestation of one's goals, dreams and desires, or direct alignment with an outcome from the inside out.

There are many dynamics to be considered around manifesting, creating and achieving one's desired state of being.

1. Visualizing your preferred outcome continuously or in intervals.
2. While envisioning your goal with the mind, then pairing it with the heart, body and nervous system to feel the sensations of what it feels like emotionally with the vision of your goal.
3. Practicing gratitude for that which you would like to receive and that which you already have.
4. Practicing forgiveness of yourself, others and all things which you may have passed judgement on or participated in.
5. Practicing self care and self love in order to feel peaceful and happy on the inside. (as within so without)
6. Seeing the event or manifestation as having already taken place... assuming or acting 'as if'.
7. Prayer to God, the Universe and/or your Higher Self for the completion of the manifestation, goal or healing.
8. Adding to or empowering the prayers by trusting with faith in your heart that God/Universe hears your prayers and they are already on your way.
9. Visualizing, Feeling, Assuming, Praying and then letting it go or releasing it to the Universe to manifest.
10. Meditating - and as a result raising your frequency vibration and alignment as you then make your request through prayerful visualization and/or spoken words.
11. Affirmations - Written then spoken clear direct influential sentences to your subconscious as Goals, Intentions, Dreams, Visions, Outcomes and Desires.

This can include spiritual mantras while visualizing goals.

12. Taking action steps in preparation to receive your goals.

13. Spiritual, Mental, Emotional, Physical energetic Healing to clear one's total self in order to more clearly project thoughts and feelings in order to manifest and receive. (Ancestral Healing, Trauma Healing. Clearing of entities. Karma Clearing work, Akashic or Shamanic Soul Healing, Life Coaching, Massage Therapy, etc.

Any or all of these can create or cause intersections through space time reality along the event horizon with that which has been ordered from the Universe by you whether knowingly or unknowingly. The intersection with our known and/or unknown manifestations is a direct form of synchronicity.

Now suppose that you have met a very special Soul traveller or a partner and there was a certain sequence of events that seemed happenstance as you recall or tell the story of your meeting. As you explore back through time you can imagine for a moment, all of the things that had to align for that meeting to take place. A missed cab, a red or green light, getting lost, an unscheduled visit to another store in a different town. Whatever it was that lead to being in the right place at the right time. And then all of the things that occurred to make all of those things occur. You can call this Synchronicity. You can also call the chance meeting, that may even have been predestined on a higher plane of existence, Serendipity.

One of the very best feelings in life is an unexpected pleasant surprise.

You may consider that the whole of reality with all of the people, places and things make up the synchronicity of one grand play interacting and intersecting with all it's parts and players.

Inner synchronicity happens when the heart and mind come into alignment. So they can be on auto pilot with many things for a person and their life. Though often we have something that we are reminiscing or ruminating about happening or trying to create and experience, and it can have many different nuances. When the heart and mind are in direct agreement about something it will come into being which is a synchronicity. The heart already knows what it wants for itself within us, for us and as us.

The mind has the tendency to jump all over the place considering different angles and outcomes. If the mind can become calm or centered enough to listen to the heart, the desired outcome will be more efficiently manifested. Sometimes we think we know what we want, need or desire and the conflicting patterns between the mind and the heart can hold us back from achieving the object of desired outcome. The heart can be blocking something just as much as the mind, though the mind is usually the one that's more out of step with our best interests. This is because it is so good at over thinking. How many times have you found yourself considering multiple different options only to find yourself arriving back at your very first choice?

Our heart of hearts wants to love and create loving projections in the purest way possible. It wants for us to enjoy our earthly experience through positive feelings and emotions. It can be holding this Great Love back when the person has experienced deep emotional hurt or life traumas which can prevent, subdue, stall or dilute what it truly knows is possible for us from within. Though, the mind and the primal ego can also have a hand in this. Truly at the deepest level we all just want to feel good and have a happy life, though it gets translated, projected and mirrored out in a myriad of ways based on the inner projections from the heart, mind and subconscious mind.

In this playground of 3D reality we become confused as to what will actually make us happy. Moving through experience to experience while searching for fulfillment in the moment. So the distractions to our Soul's Purpose and our Heart's True Desire can be many, whether short term in the moment, or long term creating detours.

These detours can be long standing depending on the seriousness of the decision made and can take us a while to get back on course. Some never recover or return from the detours. There are always opportunities to return to the path, though sometimes it's easier for the ego to become lost in the quagmire. When we feel that we've made alternate route decisions taking us off our life path, this is the opportunity to do some Soul searching and engaging in some emotional and energy healing. By becoming aware of what's happened, where you are in life and how you're thinking and feeling you are one step closer.

The next step is having the courage, focus and intention to start to heal your life and redirect your path back into alignment with you're Soul's Purpose and your Heart's Desire. In all of this be mindful of the ego and it's judgemental antics of inner negative self talk telling you it's over, you can't get back to where you once were, or you'll never be able to recover or catch back up. You have the power to break this illusion, no matter how daunting it may seem. The truth is you are in a new moment with newly inspired self healing focus and intention. The negative karma carnival is over. When you shift it inside, the Universe knows, God knows and things can now start presenting themselves to you in a better organized and more optimized way.

When we are lost in confusion, the synchronicity could now become a different parallel reality version of what's actually possible for you in the holographic Universe. As you shift your life and inner self through intention, healing and inner vision, your life manifestations start to shift into a happy, supportive and more loving version of you and your life in the holographic Universe. It's all right there for you and it has been all along in the infinity of the hologram. Then the serendipities which may not have been able to reach you before can show up in your life a little more often. Serendipities can also be described as Grace. Grace is a blessing paid forward. Maybe something that one may feel is undeserved, unexpected or that they couldn't find their way to ask for.

In the world of spirit with the Ascended Masters and the Arc Angels, synchronicity and serendipity are always occurring. Their world is very different from the duality existing here. Though, they have all come through it and understand it. They understand how synchronicity and serendipities can be great guide posts creating opening shifts in a person's life, no matter how small that shift might be. According to our Akashic Book of Life we are here to transcend, work out, and circumvent many of the things that we chose to experience and work out prior to incarnating. Some of these life events were either planned or a result of our planning what we asked to achieve or overcome through experiences while we are here. Sometimes these goals or lessons can have unexpected twists that we have to work through.

Also there are unforeseen events that can intersect with our path. Accidents or illnesses which were unplanned. World calamities or wars. Confrontations or misunderstandings which could occur. The question arises: Are we in alignment or responsible for these kinds of events? No, not really. We do participate with a wide angle perspective through the group consciousness but not individually responsible for bearing the weight of the world (unless you are a world leader or representative that is directly responsible for influencing the greater whole).

While some would say that we are absolutely in alignment with these mishaps and that we manifested them. This is done so frequently from a teaching stand point to get the listeners to take more or full responsibility for their own actions in life. Accidents and misaligned situations and events can and do happen which can be beyond our control in the moment. Our true key is how do we react or respond to them. Keeping your vibration high and clear is the best way to steer clear of a large percentage of this worldly noise.

The Ascended Masters and Arc Angels slip into 3D reality to support and guide us. To protect and sometimes rescue us during those inopportune happenings. They can guide us or nudge us in the right direction. They can bring Grace as to cause synchronous events to line up for us in our favor or to take the edge off the situation at hand while helping it to quickly dissipate. This can cause Synchronicities, Serendipities and Grace to be experienced.

The nature of the Universe is Good. The nature of Cosmic Energy emanating out from the Center of Creation is inherently Good. So ultimately we are not at the throws of a static or equally good and bad dualistic system being emanated out from God Light Source.

We are made to think and feel through generations and ages of programming that the world and reality as part of the Universe is an equally good and bad place with it's scenes and scenarios. Again, ultimately this is not true. What are we projecting on to the mirror of reality individually and collectively? Yes, the mirror is indifferent but it is projecting pure creative God force substance as it creates a myriad of projections. Yes the world and the reality is of a dual nature illustrated to us by the symbol of the Yin & Yang energy. And all things are animated at a base level with cosmic energy which is not indifferent. It is inherently good.

We develop wisdom as we perceive these natures and

begin to navigate life with them in mind. This gives us a foot up when we are perceiving ourselves in perspective to the grandness of reality and the Universe both locally and collectively. The great Masters of Light, of Love and Wisdom have figured this out long ago and they use these principles to heal, create, travel and manifest whatever they choose.

So keep your perception open. Know that you are not at the throws of a great uncaring system which is devoid of love. Those who cannot view the Universal through a connection of love, gratitude and compassion always have the opportunity to experience reality from a place of love, gratitude and compassion. Those who willingly choose not to experience the Universal through love, gratitude and compassion are then at the throws of good, bad and indifferent circumstances as well as the great reality mirror's karmic reactions. Though, they too can always have the opportunity to shift their perspective into one of love, gratitude and compassion in any moment. Forgiveness of self and others is key in making this shift. Practice being less judgemental. Practice having better thoughts, words and deeds. You may always ask for Grace and Serendipity in any situation through prayer and meditation.

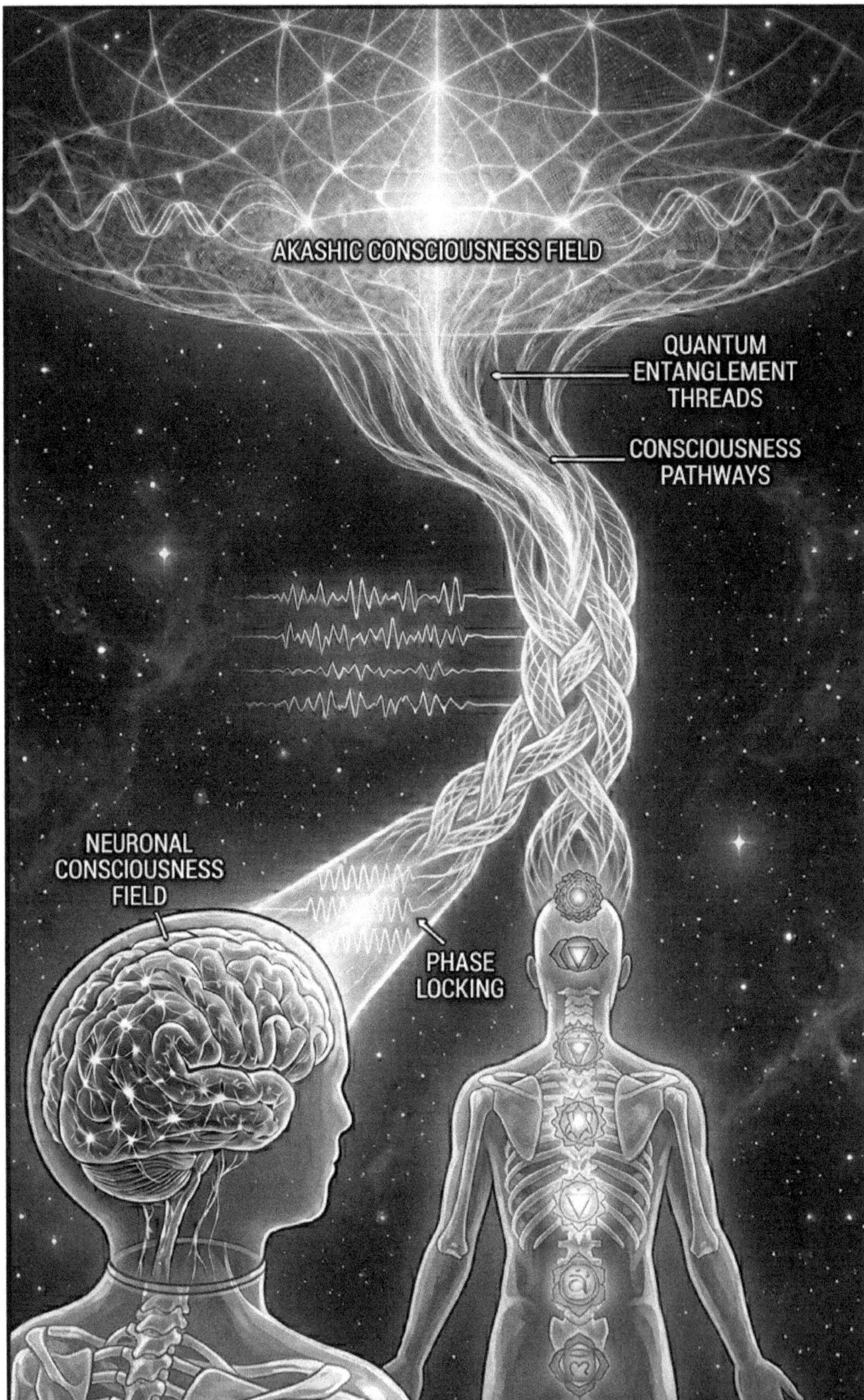
AKASHIC CONSCIOUSNESS FIELD
QUANTUM ENTANGLEMENT THREADS
CONSCIOUSNESS PATHWAYS
NEURONAL CONSCIOUSNESS FIELD
PHASE LOCKING

Chapter 11

QUANTUM ENERGY & ENTRAINMENT OF CONSCIOUSNESS

Quantum Physics - is the study of matter and energy at the most fundamental level. It aims to uncover the properties and behaviors of the very building blocks of the nature of reality and the Universe.

Entrainment - The process of one system, like a fluid, rhythm, or oscillation, pulls another into synchrony, adopting its pattern, frequency, or timing, seen in physics as fluidity flow, in biology as circadian rhythms, and psychology as brainwaves. It can mean a fluid dragging another: like air into a cloud or rhythms aligning like clocks ticking together or music influencing mood.

Begin to think of something that you like to do or something that makes you happy. Soon you will notice that your body is feeling good or you have a good feeling inside. Synchronicity of your good thoughts and feelings starts showing up in your external reality even as Serendipities. This happens when the quantum field of Universal energy responds to your vibrational frequency. Your inner vibrations and frequency start to oscillate, resonate and radiate until the outer quantum fields become entrained and respond in kind. So you get different variations of the frequency alignment and entrainment showing up continuously no matter what you are thinking and feeling. Are you getting new and improved versions of life events? Or are you getting more of the same, or maybe a standard middle ground? Or are you getting challenges that keep coming and the more you react to them through thought and feeling the more of it you get? If your outer world is calm or happy you will get more of the same. If your outer world is experiencing some sort of challenging situation or problem, either you figure out a way through it or around it, it becomes cleared up all together, or through negative reactions and attraction patterns you get more of the same.

Take any event in your life and break it down into the intersecting elements and look at it. Let's say for instance, a young man goes into a studio for a television interview. As he sits down in the chair the interviewing show host is sitting across from him. As they look at each other smiling and say hello, they both exchange their informal greetings and a comical comment is said by one of them where as they both laugh or chuckle. The production tension is broken by a moment of camaraderie. As the two gentlemen ready themselves for the interview they discuss comments on what they will talk about in the interview. They both raise a coffee cup and touch them together making a cheer of good success for the interview. As now they begin to record the filming of the interview episode.

In this movie scene there are many kinds of entrainment of consciously aware energy happening simultaneously creating the overall dynamics of synchronicity.

1. They have both agreed to meet at the studio and shown up there at the same point in time. 2. They both make eye contact 3. They both speak to each other exchanging greetings. 4. They both have a laugh together. 5. They both discuss and think through details for the interview. 6. They both raise a cup and touch them together as a toast of good vibes. 7. They do the interview together. 8. And finally, it is broadcast out to reach other people and their interactions and perceptions of it. You may take any life event and do a breakdown of it in this way to examine it.

All of these are intersections in time and space each creating a synchronicity. Whether separately or all together they are creating a dynamic of quantum energy being localized into consciously aware and entrained thought, word and deed. Creating the overall event. This is the breakdown of an event being manifested in space time reality along the event horizon. You can think of any event in your reality this way.

When we recognize the difference between singularity and the inclusion of oneness we are aligning our consciousness with the entrainment of not only these principles but the energy absolutes there within. When we are participating in a spiritual gathering and it has a very happy feel good atmosphere, it is easy for us to become entrained with the general energy of the event. It is easy for us to get pulled into the greater field of higher frequency and express along with it. Likewise if we are feeling heightened in a state of joy, love, happiness and general well being, it is easier for someone else to become entrained to our energy or lifted up into it in such a way that it raises their frequency and allows for them to also be in a state of joy, love, happiness and general well being.

Now, it is also possible to walk into a room where the tension is so thick you could cut it with a knife. We've all felt that at one point or another. You walk into a room and it goes quiet...until somehow they become comfortable with your presence and think you might be willing to participate or be a bystander of the low frequency arguments or conversation, as they start it up again. Usually if you do not become entrained to it, you will leave that setting and move on to free yourself from becoming entrained in a low vibe conversation or argument.

Quantum energy is truly a great reservoir to create from. It is able to create through us many unique variations. In an area that has a particularly high Earth energy or spiritual energetic vibration, such as being in the mountains or at the ocean, people in general tend to become more creative. They are generally happier and more agreeable as well as busy having fun. I have noticed that some folks living near the ocean really need to be living near the ocean. The entrainment they experience from the ocean's energy may take the edge off of emotional or mental issues they may be dealing with or have always lived with. It can also help people with health issues. I have also noticed that the evolving Souls can also be an open channel for quantum energy when they are in a higher frequency location more of the time. They can be prone to expressing, as an open channel, the good, bad and indifferent thoughts, words and deeds as they ride the waves of energy. In this scenario they can be very active and sometimes over active or even reactive through the heightened energy.

The Indian Vedas call this 'indulgence.' Cautioning it's students to become aware and to be present in every moment. To walk and live with virtue. To live consciously and with compassion for all beings. If you are in a coastal area and it's crowded, you may get more of a 'free for all effect,' than a 'let's be kind to one another' effect. It also depends on the energy being released from the earth grid and it's power level in any given place. We tend to think that the government or the 'deep state' is out to get us, when another scenario could be: lower evolving Souls or people living from a lower consciousness level think that others in secret are out to control them. This can be the unconscious primal self or group self in more of a reactive position than one of entrainment or alignment with higher consciousness of productivity and true success. Again what are you entraining or aligning yourself with? If you think it you may entrain with it. Live in the Tao.

Learn and practice entraining and aligning your consciousness with the Divine Consciousness. Move into agreement with all that is good. Focus on the best things in your life and the goodness that you choose to experience in your life. Clear the path for simplicity, strength, happiness, good work ethic, gratitude, forgiveness, harmony, abundance, creativity and love!

Seek to connect with your own Soul Field through the Akashic Records practiced meditations. Your True Essence is waiting for you there. As you connect with your Soul's Purpose it becomes easier to let go of the lower garbage programs that have been running through you mental and emotional bodies. When your Soul comes in and connects more within you, your Heart will come online first because it, like the Soul, it also knows the Truth of the Universal Light, Love and Pure Creative Essence. Your mental and emotional space can finally be free and clear. You can create, love and enjoy your world without offensive interruptions. Maybe those things are still there in the world but they are no longer allowed to rule over your happiness.

Practice going into your own Akashic Records. Your Book of Life or Book of Lifetimes. Each time that you practice this connection, you are signalling your Soul, your Higher Self and the Keepers of the Records that you are making progress and ready to Know the Self (Know Thyself) and they are ready to help you grow your light frequency in a way that is beyond your ego's understanding. In fact while this continual shift is taking place, you will have lighter and clearer thoughts and emotions begin to happen. Because you are now releasing the past as you continually practice stepping into the Eternal Here and Now.

Lighter and clearer living with less thoughts and emotions and without the emotional sludge gumming up an overly active ego driven mind. As you continue forward in your healing and spiritual studies, your mind will get clearer. You will be able to clearly and more effortlessly multi-task if needed or to slow down your thinking process. Many Masters have been quoted on saying, that when you start to reach levels of Self Realization you will start to have less thoughts. When your mind and heart are running on Love, Gratitude, Forgiveness and Compassion continually as a frequency vibration, there is no reason for all of the extraneous thoughts that the ego would continue to chew on.

Sometimes it happens so gradually you don't notice the shift until you check in with yourself. Others could have a more dynamic occurrence, such as the Soul flooding down into the body, or the appearance of an Angle or Master helping one to Ascend their energy in the moment. It could happen by way of a near death experience (NDE). Or during an accident or Illness. It doesn't have to be a harrowing experience. It could be an Out of Body Experience (OBE) or even an alien encounter!

Remember: 'Ask and it is given', 'Knock and the door shall be opened', 'Seek and you shall find' The instructions for asking and receiving have been given long ago and they still apply. Realization can also come in the form of a breakthrough moment when you realize all of your weaknesses or your strengths. A moment when there is an emotional crescendo. This can be a moment when you transcend your limitations and you become impervious to outside influences to overcome an illness or great challenge. Reinventing oneself in a way that cannot be denied. That Great Moment of Self Realization can also come from a moment of complete opening of the Heart. Becoming in love through bonding with another or pure and total acceptance of the Self. It may also happen as a great heart opening to the Unconditional Love coming from Heaven.

This is the pure and unfettered entrainment to a Love that is beyond the understanding through human consciousness and can only be experienced. In our world we have this happen while watching heart opening movies or attending a wedding or celebration of life. An event that allows for an opening to occur for a short time. Later the heart may balance back to it's normal state within the duality. Spiritual classes and motivational workshops can provide for this shift to happen. The masses can sustain heart openings for a shorter period of time before re-engaging in normal everyday life. In hopes that it would always be greater in numbers, fewer are those who are willing to really go into the heart and stay in it long enough for the Great Shift to happen.

While it is always available and possible, what are the circumstances and events in a persons life that keep them from experiencing the Greatest Love? The Love and Peace that passes all understanding. Often times on the Earth plane we are presented with difficult events which require a heart opening response. The transition of a loved one or pet. The end of a relationship or a difficulty in life that is insurmountable. It can also be caused by addictions to drugs or alcohol. So whether we like it or not, we all are going to be experiencing heart openings to some degree in the Earth Plane School of Existence in the forms of reactions and responses.

All the better if we start doing the work and learning to move beyond our past. To heal ourselves through learning to let go and being totally present in the Now Moment and moving forward with open eyes, an open mind and an open heart. This is the model for Self Realization or Entrainment of the Self to the Divine Love Consciousness of Heaven. Empty the mind, open the heart and look for ways to be of service to others in the world. Before you know it you will fall in love with the world and it will fall in love with you and you will wonder how it happened.

> ***"And God said love your enemy so I obeyed Him and I loved Myself"   Kahil Gibran***

Chapter 12

THE MIND EVOLUTION & THE ASCENSION OF MAN

From the time as an infant we are growing our brain and learning to survive through the care and nurturing of our parents and our surroundings. As we grow into our younger years, we start to learn and absorb traits and subconscious protocols from our world and our parents including their emotions, words and deeds. As we carry that into our teens we are now well on our way to also being influenced by classmates and teachers. As we are developing our brain early on, our mind field is also magnetized to our human form along with the emotional body, our chakras and their multi-layered fields. The mind field is the true source of all of our thoughts entering the brain. Though, in school we are taught that the brain and it's different regions are the source of our thoughts. So we continue to process our thoughts in that way. We are not really taught where the emotions are coming from, though they work in a similar fashion. The emotions (energy motions) are like thoughts, frequency based, and exist in the frequency bands of our atmosphere or the Great Field. Thought forms and emotions are then also compressed here in our atmosphere by the amount of gravity that our planet radiates. This is why we over think and over emote. In the higher less dense dimensions which are not experiencing gravity there are far less thoughts and feelings, making it much easier to relax and to be in the Great Field of Creator's Unconditional Love.

This denseness of compressed thoughts and feelings is happening constantly through us as we are a **Transceiver** or, a sender and receiver of energy frequencies. As we are sending and receiving these energy frequencies of thought and feelings we are simultaneously experiencing the hologram of the 3 dimensional reality, and further assigning our thoughts and feelings to ourselves, other people, places, things and situations. In this holographic 3D reality we are surrounded by nature, civilization, homes, streets, buildings, and all of the many creations of stuff. It's either our stuff or someone else's stuff.

1. We compartmentalize our personal (physical and situational) stuff in our life and assign thoughts and feelings to it, often generating more stuff. And then we have the others. Other Souls in human form with mental and emotional bodies like ours. But we all have different 'stuff', different personal projections through

thoughts and feelings about all this stuff. Everyone else's stuff in differences and comparisons to our stuff even though, it's the same 'stuff.'

2. This is happening while we are interacting as Transceivers through the thought and emotional frequency bands of the Great Field.

While these two dynamics are occurring simultaneously, we then also have our egos as our survival self or lower self assigning ownership to all of the thoughts and feelings around our stuff, others and their stuff, and even world stuff!

So now you can see the overall dynamics that are continually occurring. The dynamics of this are staggering and we are lost within it. Trying to navigate our way through it. And the great facade is that it's all real and it's ours (or someone else's). Thankfully there is hope, in that, more than ever before in history we have access to information about all of this and what's truly going on. As we learn about it, we assimilate truth and learn to navigate our life, reality and the world with more wisdom.

While we express all of the thoughts and feelings around everything and everyone out into our world the ego assigns ownership to these thoughts and feelings making them charged with more energy here in the compressed gravitational surface atmosphere of the planet. This is one reason why people often feel better when they live at a higher altitude such as in or around the mountains where there is less gravitational pull. Or at the sea or by a lake, as our bodies are 60-

70% water. As the Ocean air comes inland it clears the densities of the thought and emotional circulations in our energy fields and carries them out to sea. Clearing us continually. This is one major reason everyone feels so good when they get to the ocean and why it's so populated along the coasts. Some people actually need to live by the ocean because of this dynamic. Though, it is completely possible to do clearing and release work from anywhere.

So again, our thoughts and our feelings are really not our own, or rather, the frequency bands (*high to low*) that generate certain thoughts and feelings are not our own. We are entraining, aligning and processing a certain frequency bandwidth in the spectrum which we then generate thoughts and feelings from based on our personal environment. As the emotions are felt, the frequency vibration fields of them are then labelled with names, through our thought mind perception of them, further creating them and giving them form. 'I am Happy' , 'I am Sad' and so forth. As the field of vibration comes to us and through us, while causing our perception to align with the qualities of frequency as thoughts and feelings, creating personal 'ownership' of them. This gives us a 'personal' experience.

By default, we make them our own because from our perspective, it's being generated from inside of us. Now you may not want to bring that up in the middle of a heated conversation with a partner or family member, because it will only complicate things. But, this gives us a foot up when we have been under the thumb of our past thought and feelings. We are trying to find relief and are often not able to because we tend to live from our past and carry it with along with us. If we truly come to know this, we can allow the thoughts and feelings to flow through us without continually carrying them or being controlled by them.

We can take responsibility for our thoughts and feelings by knowing what we are doing with them in the moment, and learning to clear and release the lower frequency rubbish that we thought was so important to keep running through our nervous system over and over. The ego in survivalist mode is the part of us that maintains the lower egoic mental and emotional tendencies through competition and confrontation with our own selves, others and the world around us.

When we start to study the whole cyclic process that's happening, we may refer to becoming the 'Watcher on the Hill.' The ancient process of mentally projecting oneself outside of oneself in order to view the Self, surroundings and their life from a distance or outer perspective.

This simple process is that as we train a peaceful mind to be the watcher, we develop a way to perceive ourselves more readily. When we are told to simply watch our thoughts and feelings from within to try to examine them for the purpose of changing or augmenting them, we are often thinking and expressing too quickly from the ego in order to catch them in the moment.

By practicing being the 'Watcher on the Hill' we

can develop an inner protocol of watching the Self without tipping the self off mentally or emotionally. And in this little technique we can see how we are acting and reacting. In this process the unconscious primal self and the ego now know they are being watched from a higher mind perspective. It can help you in becoming less reactive to the world around you. This process is really quite profound in it's the simplicity and can bring you newfound perceptions and ideas about yourself. There are a multitude of practices for letting go of thoughts and emotions. I will share here a couple powerful examples and then a few releasing techniques and an even more powerful dynamic happening behind the holographic scene.

Letting go of emotions within certain areas of the body can be very helpful. If we look at the Vagus nerve system that runs from the brain through the central core of our torso touching each of the organs we come into contact deep within ourselves with the very center of where it all happens.

As we look at the mind field encompassing the physical body which is entrained with the brain we also see the top half of the equation. The mind field is connected to a multi-layered field outside of us that magnetically hovers in the atmosphere of the planet but also not limited to only the atmosphere. It can and does connect us together between individuals and groups in our world, and also connects us with off world sources such as our Guides, Higher Selves, the Angels, the Ascended Masters and God. As we connect to others and our non-physical community, we are syncing with thought form variations on multiple levels of frequency.

As we then attract into our fields variations of thoughts that can be used for various and specific processing and tasks. The thoughts enter into our brain on different levels and in different ways. The thought field vibrations can enter through our crown and through our 3rd eye or brow center. Thoughts, images, pictures, visions and words. Thoughts also continually enter in from our mind field through our ***eyes***, whether they are re-circulating or stored thought forms in the mind field or they are newly introduced into our field from the Great Field. This was known and studied in ancient India by the Ayurvedic masters who had also solved the mysteries of quantum energy, mapping of the stars and revolution of our galaxy around the Great Central Sun. In fact, their approach to advanced levels of longevity and mastership were born out of their want, need and drive to continue their metaphysical and scientific studies.

As the thoughts enter into the brain through these channels, we then formulate thoughts inside the brain which are processed as we assigned different levels and associations of importance to our own personal self, based on past, present and future moments of consideration. Everything from the ego's colorful charades based on survival, comparison, competition and confrontation, to the more balanced interdependent creative and loving associations with self, family and society, and even on to the greater associations with the world, civilization, Greater Self, Universe and God. As we know, the human brain is a very busy place, processing anywhere from 12,000 to 70,000 thoughts per day both voluntarily and involuntarily. And then the Great Field is, as you can imagine, filled with endless thought form and frequency variations.

As the frequency attraction or attractor patterns begin, the thoughts move through the mind as they are formulated and drawn down through the Vagus Nerve which is housing an amazing 4 fold system.

1. ***The Internal Body maintenance function*** - The light speed messaging (high speed modem) to all parts of the physical body for functioning, balance and all nutrients, endorphins and peptides.
2. ***The Dual Operating System function - switching between*** - Para-Sympathetic nervous system function Feelings of comfort, joy and goodness. And the Sympathetic nervous system function of 'flight or flight' - anxiety, stress, fear, panic, terror, etc.
3. ***The Subconscious Mind function*** - projecting/ broadcasting a signal to manifest into reality what is intended through the Self, out into the fields and to the local reality what it is being given by the conscious mind, the ego, the inner child, or the unconscious primal self.
4. ***The Emotions function*** - Energy motion impulses generated in the Vagus nerve and used in league with the thought impulses and emotional field. Feelings that are felt in the body tissue, the heart

and the mind. Feelings are known to increase the function of the subconscious mind emanations.

The emotions (energy motions) must be felt. And we feel emotions in the body by way of them coming through the nerve impulses via the Vagus nerve. They are then felt in the Brain and Heart and radiated out into the physical body tissue, the layered energy field around the body and into the atmospheric field of reality in unison with the subconscious mind and it's broadcast field. The layered signal emotional and subconscious broadcast frequencies are then colored by our existing reality, the ego, Higher Self, the Universe or God as these thought/emotion frequencies are simultaneously broadcasted out into the reality mirror.

The Vagus nerve acts like an ***inner wireless wifi antenna***. Radiating out a field of vibration around the body and then further out into the Great Field, which in turn manifests our local reality, as well as situations, people, places and objects. The Vagus nerve is both hard wired into your physical systems and a wireless broadcasting system of it's own multifunctional purposing. (You may want to re-read these 2 pages so you get a working model of the energy flow of thoughts in from the field, charged by feelings, then radiated from the subconscious function back out into the Great Field).

As this wireless and hard wired broadcast operates continually within and without, the energy of the subconscious protocols and feelings can become stored in our body tissue as well as the energy fields. As we experience a new situational moment that resonates with past thoughts, feelings and situations, the emotions and thoughts are once again called up from the tissue. This can bring an intensified or prolonged version of the mental/emotional state whether beneficial or non-beneficial. This is why so many of us truly have trouble letting go of old stuck thoughts and feelings about our past. Your whole system may be looping on auto pilot.

Now that we understand this intricate system within us for processing the thoughts and feelings, lets talk about the dynamics of clearing, releasing and letting them go. The focusing of conscious awareness through the mind focused down into body on a particularly uncomfortable emotional charge or area has the power to release the emotional discomfort or charge of energy.

In this procedure try not to focus on any colorful techniques or processes. Simply clear the thoughts by giving the brain and mind the job in the moment of holding your conscious awareness on the uncomfortable feelings or sensations in the body. As you become aware of the painful area, go into it, be with it, focus into it and through it and breathe. Then as the conscious awareness is being present with it, just start to let it unravel and release from the tissue. Very simple, no super involved formulas of healing or visualization, only the conscious mind focusing it's awareness to that part of the body.

Let it go, let it unpack, and let it leave. You can hold your focus on this continually for a few minutes.

As thoughts about the origin of the physical pain or emotional anxiety come up, simply let them pass and let them go. Give the mind the job and focus of, 'right here and now,' being present with the area of discomfort or discontent and continue to breathe.

So for instance, if you are releasing old layers of fear buried in the tissue, you can use fearful traumatic memories to activate the stored emotional energy in the body. Once you have the emotional charge activated then release the thoughts and keep the mind focused on the emotions and wherever they are in the body. As you bring them up stronger, eventually the charge will run out naturally. It may be quick or it may take a while, but it will happen. Generally you can move a lot of old emotional stuck energies out in 20-30 minutes at a time. Some things can be released very quickly and some things may take longer, but if you will give it a try and stay with it, you can start to experience relief, feeling lighter or no pain. This is the seemingly miraculous power and affect of the pure conscious awareness being focused through the mind and down into a location of the body.

This has been exhibited in many ancient spiritual techniques from the masters, gurus and healers of India. They could project consciousness through there minds or their eyes and heal others. Some would say that it's because their energy is so high, which is also true. But it is also the ability to empty the mind of trying to do anything and focus pure conscious awareness to another person from their consciousness lense of the mind. It is done by using the mind to funnel in the consciousness without focusing on the mechanical thought process. Focusing 'their' consciousness on the person or part of the body to be restored. Also this is effective because they are not personally connected to the individual's back story.

In modern times, the Science of Mind studies that were based in Los Angeles would call this a 'mind treatment' You can also ask your spiritual helpers or the Great Field to help release it for you as you give it up and release or let go of it, whatever it is, from within. Keep the process simple though, as that seems to help greatly.

Guatama Sidhartha (the Buddha) recognized this in another way of not only holding the focus in a certain area for a prolonged time, but also scanning the body with the mind from head to toes and back up again over a prolonged period of time. This became known as Vipassana. It not only releases pain from the body or trapped old layers of emotional discomfort but also life, body, mental and emotional karmas and patterns that may have been looping in a person's total being or their life indefinitely.

The result can be lasting peace and balance of emotions and thoughts if practiced continually and in long enough intervals for change to be felt. This technique can be learned through a Vipassana center or from someone who knows the techniques. The aforementioned process of letting go through breathing and mind focus is easily practiced and has literally helped thousands of individuals regain balance by releasing past traumas. Releasing body ailments has also been achieved. Though, if you need to see a doctor you should never stop yourself from seeking medical care. Letting go can be a helpful prevention or even a supplemental recovery technique after medical treatment.

Insights on Levels of Ascension

Since ancient times the techniques and practice of finding one's way into the afterlife or back into the 'real world', where there is no death has been a sacred ritual and practice among the high priests, sages, siddhas, shamans and mystics. Whether in a communal study within a civilization or individually on their own. The nature of consciously being connected to a higher realm, the stars, or a sacred group of divine beings in the Great Beyond has held as a special place of practice in readying oneself for their end of life journey.

Many spiritual lineages, tribes, groups and individuals have even practiced the art of ascending their total self (including the physical body) earlier than their end of lifetime transitional point. Changing their Soul schedule and life path in unique ways. As an Indian sage, the young Siddha that became an immortal master is Babaji Nagaraj. Through tantric Kriya yogic techniques he brought his Soul Body down into his teenage human form and transformed his life span beyond that of the normal aging process. This revered Indian saint is said to be over 2,000 years old...

or young, depending on how you look at it.

In this heightened state, his Soul essence brought a unique dynamic which allowed him to experience all of the siddhis (special powers) that transcend time and space. He frequently appears to students and seekers of the Divine around the world and in the area of India with messages, healing or manifestations in order to help those who had followed his teachings of yoga in a past life or someone who simply needs help to receive a blessing that may be beneficial to their path.

There are many other masters in the Indian culture who have achieved not only enlightenment or self realization, but also certain levels of ascension. Masters who do not eat, some that don't even drink or sleep while maintaining a healthy physical body on the Earth plane. For all of them that are known, there are many multitudes of others who live at such a high vibration that they live outside the reaches of society and are unknown to the world. Moving in and out of physical form as they commune with nature itself timeless in their essence of being.

Jesus the Christ, as we know Him, in our religious cultures around the world, also experienced ascension. His was a very unique story that affected all of humanity both directly and indirectly. As he studied God and Creation from a young age it wasn't long before he was naturally teaching and witnessing to people of all ages. As he was born from a unique group that existed outside of the Essenes. Mary and her family had practiced a devout communing with nature and had also achieved immaculate conception for many generations. This is from the Records and outside of the normal known story line, though there have always been those who have known of this.

Jesus, Jeshua ben Joseph, Yeshua or Isah, as he has been known by many names, travelled to study abroad in the 'lost years' and be received by many mystics, masters and high priests throughout Asia and the Middle East including India, Greece and Egypt. They all acknowledged his path as a sacred way shower and teacher and shared with him many of the insights and energy tantras, internal healing arts and mystical secrets reserved for the higher adepts of their cultures.

As he came back to his homeland region he had become naturally enlightened and self realized and in harmony with the Great Over Soul of Creator God/ Creation/Universe. As it started to pour down through him, he continued to teach and heal the masses. He had both the consciousness of God and man within him. As he was captured and went to be crucified the great opening that he endured during the torturous event created an opening within Him. His DNA which had God consciousness flowing through it became opened and flowed out affecting or energetically touching all of humankind through the holographic reality. This was not known at that time. (again from the Records) A great opening occurred as the God Light was released from his DNA that influenced the generations of humanity around the world and to those who would follow his example and teachings of

Love, and to the whole of humanity en mass as a silent blessing for progress through love. A great shift in human consciousness was started like never before in a very unique way as it was unleashed from his body.

His path was predestined by God for this unique dynamic event to occur. Even though he knew it was coming, the human side of Him did experience fear, even though he went forth straight away through courage. All that witnessed this and even all of the Roman Empire after it was done knew that an ominous act had been committed. They could feel it. It shifted them as they now had compassion and even regret for what they had done.

As he was on the tree, he went into a deep state of sammadhi, an Indian master technique for leaving the body and putting the physical body in a state of suspended animation by shutting down all bodily functions. After the burial, his body was retrieved and revived by his close associates and family members, as they made there way out of the country. In that time period during the crucifixion and after the burial he was reported to have appeared to many of his friends, family and followers. He also was reported to have appeared across the globe in distant regions, with energy blessings and messages. He was truly in a suspended state of ascension during this time. He had achieved what he set out to do. By being turned into a martyr in the name of God's love, the event, story and energy dynamic of it all touched humanity in a way never seen before or since on our planet.

As the catholic institution arose out of the Roman empire, Jesus was at the head of their crusade and the Holy Bible was compiled in different stages. It was such a tremendous event that it really inspired the catholic movement, even though many believed that they were capitalizing on it. Throughout the following centuries nuns, priests and other deeply spiritual people across Europe and beyond experienced what is known as **stigmata,** or the wounds of Christ. These individuals would often bleed continuously for no reason from certain areas of the body where Jesus was wounded such as the hands and feet or their side, and they also often exhibited spiritual healing powers and other mystical gifts. These were usually people of great faith or connection to the Master. Another miraculous manifestation of the holographic Universe.

So the DNA of one man, *The Man,* cracked open while God was flooding through his body and touched humanity through their DNA inter-dimensionally in the holographic Universe in ways which are beyond our understanding. This is an amazing incremental (slow motion) mass ascension for all of humanity in a completely new and different way. Now the common lay person everywhere had the gift of studying love, forgiveness, and many other virtues openly in an age where it was much needed. It was a turning point in our mental, emotional and spiritual evolution. Before that people were governed by the laws of the church officials. After that, they had a new way of looking at what really mattered through freedom, love and virtue.

There are masters in most every culture from around the globe that have experienced ascension either in the traditional way at the end of an incarnation or during their life. The alchemists of the medieval and renaissance periods of Europe were said to have achieved immortality through the creation of the 'philosopher's stone'. One such example was Count Saint Germain. Other's had used the ancient Egyptian and Indian techniques of Solar Gazing which involves drinking in the cosmic rays of the sun at dawn and at dusk through their eyes. This produces an oil called chrisum that is excreted from the pineal gland, allowing the participant to virtually live on light with or without consuming food for sustenance.

Eating healthy,fasting and doing regular cleansing procedures for the body along with yoga and meditation are sure ways to help one achieve a high vibration which is a precursor to ascension. Also by releasing old thought patterns and belief systems, as well as releasing old emotions and traumas from the self in body and spirit are helpful and required in raising the vibrations of one's self on the journey towards ascension and a healthy happy life.

As we have talked about in an earlier chapter almost everyone who has ever had a NDE (near death experience) would prefer not to have returned from the divine heavenly experience. When this happens they are in the Soul energy completely. I have found that by way of practicing the accessing of the Akashic

Records and your own Book of Life within the Records this activates your Soul Body which surrounds or hovers over the human body. Accessing the Akashic Records is one of the only ways besides advanced tantric techniques that I've found that simultaneously activate the Soul body's energy awareness and it's connection to your chakras, mind, senses, nervous system and multi-layered energy bodies including the subtle body and the astral body.

As you continue the practice you can have a natural state of enlightenment or self realization of the True Self occur. You can also experience what I like to call a mildly extended continuing NDE or NLD (Near Life Experience) occur for you ongoing as it incrementally builds. This is more of a homogenized version of ascension that's safe and doesn't have the over the top drama involved with mishaps as a way to see heaven. You are willingly entering into that state through continued practice of journey meditations.

Also, if there are any events in your Book of Life that would point to an NDE this can be circumvented by accessing the Akashic Records as a study practice. Not to down play any NDE that you may have had in the past because they are very spiritually Divine events that are life changing. If you have previously had an NDE then you may find it very second nature to visit the Akashic Records as a spiritual practice and receive additional insights and activate new mystical gifts!

The ascension of the Earth is also something to touch on here. While we as Souls are individuated we are all connected as part of the holographic Universe. The Universe, the dimensions and the time lines are all evolving. There are different realities which are also intermingling with and influencing our current primary reality time line. One story which is more of a simple interpretation of this, is the '5D reality' concept which promises that we will be shifting into the 5th dimension. It is better stated that the 3rd dimension will accommodate levels of 4th and 5th dimension bringing a much more diverse outcome if indeed it does play out in that way. Quite possibly the Earth with the rest of the planets in our solar system will be ascending through a great ascension shift. This could happen as a result of an astrological shift very quickly or over time.

As many scientists and spiritual students have acknowledged we are passing into and through a Great Galactic Photon beam of Cosmic Light that is broadcasted continually from the Great Central Sun. This is part of the Great Cosmic Procession of our galaxy's 26,000 year rotational cycle around the Central Sun. 13,000 years into darkness/separation, which we are just coming out of and into 13,000 years of Unity consciousness with a more heightened and higher vibration here on our planet and beyond. A grand cosmic season. All of these variations we've discussed point to ascension. How it exactly will occur is partially up to us as we are writing our civilization's group conscious story within the planetary Akashic Records. It will be also influenced by intergalactic, solar, and planetary physical and cosmic influences, including our Mother Earth's changes.

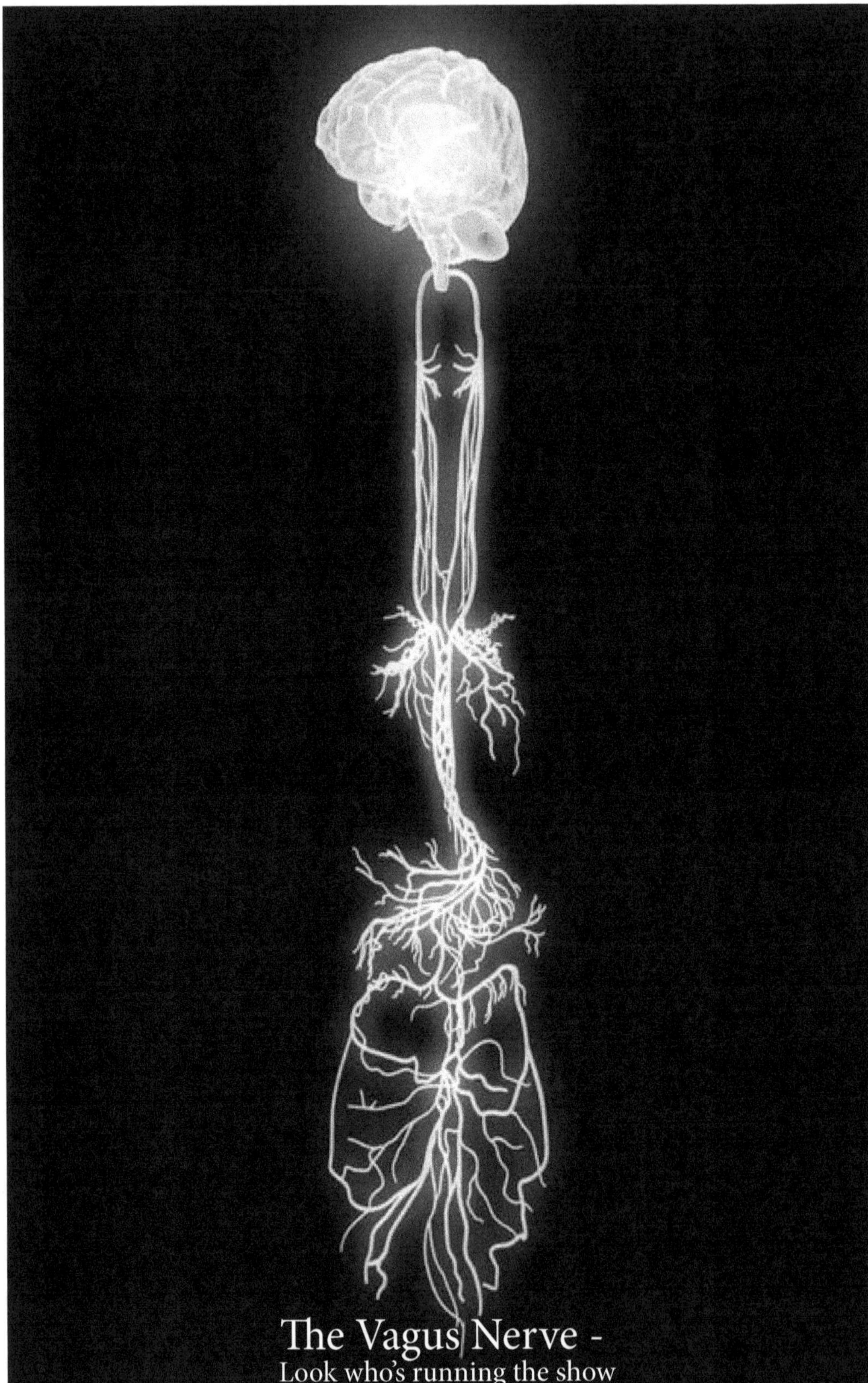

The Vagus Nerve -
Look who's running the show

Meditation Exercise 6

REPROGRAMMING THE VAGUS NERVE SYSTEM

Guided Energy Exercise - Always remember to never do this work while driving a vehicle or using any type of machinery. You must always be stationary in a quiet state. Read the script and then close your eyes and visualize while doing the exercise.

As we talked about in the last chapter the Vagus nerve is central to emotional feelings and subconscious activity. In this exercise we will release negative or debilitating thoughts and feelings from the nervous system that can cause trauma and all levels of tension, anxiety and stress in our daily lives. This is known as 'Fight or Flight' mode and we will release it here. As well as cultivate clearing with the subconscious mind in general and around any events in life that you would like to change.

While in a relaxed position place the palm of one hand

upon your solar plexus and the other palm resting upon the center of your chest. Focus your mind down now into your center and connect with your Vagus nerve as if you are quietly saying hello. Keep your focus there on it. As you breathe in and out gently we are going to do 3 affirmation commands to the Subconscious, the Parasympathetic nervous system and the Vagus Nerve.

For the Vagus Nerve -
As you breathe in while focusing your consciousness down into the Vagus nerve say to your Vagus nerve aloud in a positive tone:
"I Am Resetting My Vagus Nerve" and as you breathe out release old tensions from this inner central corridor.
Repeat this 3 times.

For the Parasympathetic Nervous system:
As you breathe in while focusing your consciousness down into the Vagus nerve say to your Vagus nerve aloud in a positive tone:
"I Am Resetting My Parasympathetic Nervous system" and as you breathe out release old tensions from this inner central corridor.
Repeat this 3 times.

For the Subconscious Mind:
As you breathe in while focusing your consciousness down into the Vagus nerve and Subconscious mind say to your Vagus nerve aloud in a positive tone:
"I Am Resetting My Subconscious Mind" and as you breathe out release old tensions from this inner central corridor. Repeat this 3 times.

Releasing and Manifesting with the Subconscious

Now think of a situation in life that you would like to clear up or change. Maybe you want to visualize a new outcome in life. Is it Health, Financial, Legal, or Relationship?

Focus your mind now down into your center and connect with your Vagus nerve and Subconscious Mind as if you are quietly saying hello. Keep your focus there on it.

As you breathe in and out gently, breathe healing energy in through your hands and send it into the Vagus nerve and Subconscious Mind. As you breathe in and out focus on what you would like to clear or manifest.

As you exhale see your Vagus nerve radiating the signal and frequency out to the reality and the Universe of what you would like to clear or create. Do this 3 times or repeatedly until you feel you have established a clear strong connection with your Vagus and Subconscious and what you are intending or releasing.

Now relax your hands and give thanks to your Vagus nerve system and your Subconscious Mind for listening and working with you.

Repeat this as needed for trauma release, getting through a difficult time in life, or for manifesting. This may be repeated as often as you like for physical and energetic comfort. Enjoy!

Mediation Exercise 7

LETTING GO OF NEGATIVE EMOTIONS AND THE PAST

Guided Energy Exercise - Always remember to never do this work while driving a vehicle or using any type of machinery. You must always be stationary in a quiet state. Read the script and then eyes open or closed and emotionally engage while doing the exercise.

Begin by getting comfortable in your favorite chair or by lying down. As you take a few relaxing breaths, breathing in and out slowly. Get in touch with you heart your mind and your inner being as you continue to relax down even deeper. Notice if you have uncomfortable feelings in your body. Are they linked to any negative beliefs? Are they linked to any past memories of situations you wished you could stop repeating in your mind and emotions? Instead of trying to turn them off, let the thoughts bring up the feelings in your body associated with them.

Really let the feelings come up within you. Instead

of keeping them bottled up and feeling it is unsafe to experience them. When you have the feeling present stay with it and see if you can feel it more. If your mind starts running endless thoughts and scenarios just direct your mind now down to the part of your body where the feelings are. Usually the stored emotions will be somewhere in your torso, midsection or lower gut. Are you feeling fear or anger? Resentment or jealousy? Pride or apathy? Hopelessness or depression? As you focus your mind down to that part of the body, focus your mind on the feeling that's in the body. You may use associated thoughts to trigger the feeling but then go into the feeling itself and stay with the negative sensations in the body as you breathe.

Place your hands on that part of your body. Is it your chest or your stomach or your lower belly? As you stay present with the emotion you are letting up to the surface, let it unpack. Often these stuck feelings are shut away and bottled up in our tissue. These are lower frequency response patterns that were generated and projected through the body by the nervous system.

As you breathe into the feeling or emotion and feeling in your body more fully, and now as you exhale see if you can just gently let go of the feeling. When you let go of the feeling completely, the associated thought patterns will stop running. All of the fearful 'what if's' will dissolve. All of the anger, fear, pain or other negative sensations are ready to leave. They are free now to be released back into the Great Field where they came from. It may take a little while so stay with it. These old stuck emotional patterns are in layers and they've been layering up for quite some time.

We had attracted these frequencies to run through our nervous system paired by personalized thoughts. We internalized them and made them personal to our life and situations. They are not really yours. They are external frequency patterns that you have invited in unknowingly. Let go of them now and give them back to the Great Field. Give them up. Let them run their intensity until they start to fade. They will lessen in intensity until they finally run out. As you continue to breathe them out of your body, you may summon them to leave by raising your hands slowly away from your body.

As the emotions come up sit with them and be present with them. It can be uncomfortable and this is the part of the work that no one really wants to do. But you do because you want them to leave your body. These old stored emotions can attract more of their selves through new scenarios that could keep us stuck. That's why we let them go.

We are done with this kind of feeling and the thought patterns associated with them. Just keep breathing and keep letting them come up and...Let it go....Breathe and.... Let it Go.....Breathe and Let it go.....Breathe and.....Let it Go. Slowly unpacking and letting them leave.

As we do this we are making room for peace and harmony. Better thoughts and feelings of relief and forgiveness. Gratitude, love and joy are waiting to come in. No more struggle in life with the old patterns. No more suffering in life with the old thoughts and feelings.

They will no longer create your future.

You may want to stay present to this exercise and run it for a few days. More old patterns will come up to be released. There is always more to let go and to release.

You can also use this exercise to release any resistance you may have around things you need to get done. Goals can be achieved much quicker when these invisible and stuck 'counter intentions' to your successful living are finally let go of. Counter intentions are intentions we never signed up for and they can hold you back through resistance. You may find that one layer or old pattern such as limiting self beliefs or uncertainty has kept you stuck on multiple projects or on getting your 'To Do' list accomplished.

Say: "Thank You God, Universe, Holy Spirit and Guides for assisting me as I consciously let go and release these old emotions. I give them to you. Giving them up to the Universe and back to the Great Field. Om Peace Amen.

Good work, now write it down. What you are feeling in the body, where it was, and what's the original situation that triggered these emotions. You may find that they go way back in your life. Be diligent and present with this work and keep letting the old emotional patterns go. You are making room for new ways of being and new experiences coming into your life. Keep letting it go, simple and persistent. As you let go of these 'counter intentions' they will no longer be blocking your true intentions for what you would like to have in life. Good Work!

Mediation Exercise 8

Body Scanning for Healing & Clearing Karmas

Guided Energy Exercise - Always remember to never do this work while driving a vehicle or using any type of machinery. You must always be stationary in a quiet state. Read the script and then close your eyes and visualize while doing the exercise.

This is from Buddha's original process that he found while sitting under the banyan tree for 40 days. After that time of his enlightenment Siddhartha Guatama spent the rest of his life teaching hundreds of thousands of individuals how to do this process.

Now known to the world as Vipassana. The long process is very opening and should be done under the guidance and direction of a teacher or ashram. There, they hold classes worldwide continually and they are easy to find, if interested, for 3 days, 10 days or 40 days. My original experience was a 10 day meditation retreat. In that process there was no speaking or eye contact with any of the other participants. Just eating, sleeping, walking and meditating.

This is what I believe to be the original process that later popularized techniques were based on. Whether they referenced the original Buddha's technique consciously or maybe unconsciously, the process of allowing consciousness to do the work by flowing it from the mind directly into the body is the same. They both have one thing in common that embodies the same dynamic. The focusing of consciousness into or throughout the body in ways that allow for tension, pain, negative emotions and karmic energies stored in the body tissue and it's energy centers to dissipate and dissolve.

This is not the whole elongated multi-day process, but rather, a distilled version of that work, that is effective, so you can get another internal perspective of how pure consciousness can influence the body. Where as, in the Letting Go technique, we focused on emotions and their feelings in a certain part of the body, this is the process of continually scanning the body from head to toe for a period of time that will also produce a full body clearing effect of negative trapped or stored energies. Including old mental and emotional patterns and even karmas.

As you focus the mind on the process of visualizing and scanning the whole body slowly, you are giving the mind a focus. You are giving it a job so it is not wandering aimlessly from thought to thought. Thinking things like 'Will this work?' 'Am I doing it right?' 'Is anything happening?' 'What am I going to have for lunch?' etc. All the normal wandering thoughts that are going to happen during meditation can be cancelled as you focus on continually moving the body through the mind.

Now, you will have thoughts come up associated with feelings and experiences in certain parts of the body and that is normal. Or if you have thoughts come up in general, that may also happen. You want to discourage the thought process by focusing the mind more diligently on the scanning exercise. If thoughts happen, just go back to focusing on the scanning process. You may not feel anything happening during the process, but after it you will feel lighter as a result as you start to move through your day. Be diligent, grasshopper.

Here is the Scanning Process:

As you sit in a chair upright with your feet flat on the ground, relax and begin to position your posture perfectly upright with no hunching. Try not to slouch or sit crooked. Hands are facing down or up and relaxed on your thighs.

Start by taking a few relaxed breaths in and out through the nostrils. As you do this become aware of the wind that the breath creates on the rims of your nostrils. Become aware of your breath continually as you focus on the rims of your nose. There is only the breath. The breath inside you and the breath outside you. Do this for 5 minutes. Focus, Focus, Focus on the nose and the breath wind sensation.

Now, as you continue to breathe naturally move your focus up to a ball of white light sitting directly on top of your head. Do this for a few minutes.

Now, start moving visual focus down around your head, mapping or scanning the outside of your head. See if you can use your left and right brain to scan in multiple directions at once. You are scanning from the top of the head down and over the surface in all directions at once. Then come down to the neck and scan all the round sides of your neck down to your shoulders, left and right shoulders front and back. Scanning or mapping the complete body surface in all directions causes your brain to engage both left and right hemispheres.

Move now down around the left and right arms at the same time. Becoming aware of the shape and details of the total surface of your left and right arms at the same time as you are mapping. Scan down to the forearms, then wrists and then both hands all the way down around each finger to the finger tips. Once you get to your finger tips, start moving back up again through the same process, scanning both arms at the same time.

Once you get to the shoulders, scan down around your torso, front back and sides at the same time mapping the complete midsection. Then down to your trunk front, back and sides at the same time. Now, just like the arms, branch off to scan each leg in it;s circumference at the same time. Noticing all the details and awareness of your body. Scanning all the way around both ankles and then down around each foot at the same time, down around all toes simultaneously and then to the tips of the toes.

Once there at the tips of the toes, return scanning

back up the body in the same way. Both legs, your trunk, torso front, back and sides. Up to the shoulders. The arms are optional, if you want to scan the arms again you can, but not necessary. Now the shoulders, the neck and then back up around all sides of your head to the very top.

This is one complete body scan. Repeat this process 3 times. You may experiment with the speed of your scan. I like to do it slowly and be more present with the detail of the body. After 3 scans you may open your eyes and start to move the body. Get up and move or walk around slowly to see how you feel.

This is how the pure consciousness can simply focus the mind on the body and begin to heal and clear it through pure consciousness. The act of the scan itself is what to focus on. The consciousness is doing it's work in the background. Do not focus on the consciousness clearing or healing the body just allow it to occur. Well done!

Chapter 13

Angelic Influence on the Earth Plane

It was a hot muggy summer day in Texas. The sun was just a dot. The kind of day you read about in an old drug store paper back cowboy novel. I was having a challenge down in Austin town. Seems there were entities attached to someone I had been working with. It was an energetically unpleasant feeling as they had shown up in my dreams a few times for a show down. As I fought with these energies in the dream state like an angelic warrior, I always came out on top, but the whole thing had gotten my attention. In the early days I wasn't doing as much energy clearing as I came to understand was necessary in the long run. Being of intuitive service to a wide range of folks, I became very sensitive to a wide range of energies. So I asked a friend for help. He said "Yes, why don't you come by and I will help clear you before you drive out of town this evening."

As we sat down together he closed his eyes and became quiet for a while, as he said," The Arc Angels want to help you clear this. Will you accept their help?"

For reasons beyond my earthly incarnation, I had stopped working with the Arc Angels. There was a time when I was with Gabriel and I saw the way deals were being made in order to get things done. Yes there are bureaucratic workings even in the higher realms. To help and work with people and Souls in a wide array of situations. I didn't agree with the way they were doing business. Though, in the higher realms there is a less distinct line which separates the seemingly dark from the light and that separation grows less distinct the higher you go until there is only Oneness and light. Once I was reminded of it, I remembered it vividly. I had expressed that I didn't agree with the way they were doing things and that I was leaving.

Now on the Earth plane again, when I had thought back to it, there was a fair amount of anger and discord around the whole thing. So getting real clear in the moment, needing the help and knowing that the Arc Angels were the ones who could do it, I said "Yes." So the clearing began as he summoned a large group of the Arc Angels by name. As they circled around me and I watched this through my intuitive Akashic Vision, it was a very unique meeting and reunification. There was a lot of love and acceptance coming from them. And equally important, I felt the dark energy that was bothering me immediately leave. I agreed to work with the angels and to befriend them again.

After this clearing session, I said my good byes to my friend and got ready in the late afternoon to leave town. On my way out, I stopped at a swimming pool for a quick swim. At that time, Austin had public pools in many of it's neighborhoods. No one was there. I remember that as I floated on my back in the water after the clearing, that I had never felt that kind of peace in my life. I felt a great peace and harmony to be reunited with my brothers and sisters from the Angelic Realm.

Since that time, I have worked with the Arc Angels achieving great results in my own life and the lives of countless others as clients in readings, healing sessions, clearings and life path guidance. They can be powerful Divine allies who are always guiding and directing humanity in our search for our true path and nature. The challenge that many people face in hearing of them or deciding whether to except help from them, is that we have learned of their existence through religions from around the world over the past 2 centuries. People often balk at the iconoclasts of organized religion, even though they may have been deeply associated with one or more of these religions in past lives. This has also included the idea of working with Jesus, Mother Mary, Mary Magdalene and others, such as the angels, saints or disciples of Christianity and other religions.

The fact is that all of the aforementioned folks were and are still connected to God, the Universe and the Earth plane and can all be powerful helpers when needed. When you need help, you need it, bottom line. I have worked with people who were under attack from entities, demons, dark aliens, inter dimensional beings, etc. that were in danger and needed help. Often times due to their own karma, fears, superstitions, health or life situations people can either attract these beings or be under the influence of them.

Everything in the Universe is energy. And we, as spiritual beings have power and energy that we are not even aware of. So in this case, these beings could be, but should not be drawing power from a person. Calling in an Arc Angel such as Michael, known for protection and justice, can be a life changing event for such a person. They have ways of working beyond our known means of working which can be very helpful.

Ultimately, calling on Jesus, who has power over all darkness can also be super clearing, super healing and super relieving. What sets Jesus apart, if you're asking masters, angels or beings for help is that His Celestial Soul is a great Universal Soul that is part of the Soul we know of as Creator God. His Soul is a great guiding power, so much that He is recognized by off world alien groups. This is why he has direct leadership with the Angelic Realm. We as humans sometimes get tripped up on the idea of asking Jesus, Mary, Michael or the angels for help because of the varied misuse of messages coming through others in organized religious groups about them. Remember that Jesus, the Arc Angels and all of the masters are beyond religion and live in the Great Light. They are always available and present. Just because you ask them for help or they come to your rescue for healing, clearing and protection does not mean that you are then karmically bound to them. They do not operate in that way. But, for those who have needed help and have asked for it from them, I can tell you that their are no greater allies. Your guides and other spiritual deities can also be of great help.

The Indian culture has Shiva, Durga, Ganesha and a lot of other spiritual deities and masters who are all aspects of the Godhead and can be very helpful, clearing and protective. Remember that they all work together and know each other on the other side so you are not doing yourself a disservice by being open to working with more than one of them. Learning to pray can be a powerful communication and connection to them. They are here to help you on your path of development in your life on the Earth plane and your Soul evolution. They are not here to control you but to guide, heal and help. They are the ones who have gone before us and they know what you are going through and for this reason can be great healers and way showers.

I primarily work with Arc Angels Michael, Gabriel and Raphael, though I have worked with many of them over the years. They all have very unique aspects and in some ways can do some of the same things for you as they all hale from the Angelic Realm. They are all on call from the Godhead to the Earth plane to keep it from being over run by the circus of darkness while we are here now numbering in the billions to evolve, learn about creating and learning to love. And they also know that while we are Divine beings as Souls while in heaven, that on the Earth plane with a human body and ego, we can all be like a bunch of unruly youngsters on the playground.

Those in heaven know how brave you are as a Soul for being valiant enough to come down and to participate in the super compressed emotional, mental and physical experiences here. Michael is known for his Great Protection and Justice, a great warrior for Truth and all that is right. Gabriel heralds the Great Light and Love of Creation and of God and brings it to us here. Raphael is the Celestial Healer and can bring Energy Healing, Physical Healing and Love. He is a very gentle and compassionate angel. His energy is emanating the Christ healing vibration in his own unique way. Metatron is the king of the angels, and while not as well known as Michael, is also a powerful warrior with Great Love. I have met many people who work specifically with him. Jophiel has a very friendly energy and is known for being inspirational with creativity. Ariel works more with women and also lovers in her growing of the presence of the Divine Feminine. Chamuel is also a great healer. They all can do a lot of the same things, though each one with great power has their own personality and variation of energy.

Dr. David Hawkins, an author and spiritual teacher, who became expanded through his own NDE's, had met the angels in some of his more expanded and out of body experiences. While being extremely sensitive and also involved with muscle testing, he had developed a calibration scale for the vibratory level of humanity and the energy of mental/emotional frequency energies in the Great Field. The Hawkins calibration scale of 1-1000 squared to the 10th power placed many masters from around the world in the 800-1000 range. He noted that upon testing through calibration, that the Arc Angels calibrated at 15,000. That gives you some idea of who we are dealing with. Beings of Great Light and Power.

Upon being in the presence of Arc Angels and witnessing their great power he expressed that it is important to know this and to be careful of calling on too many of them in at the same time, and I agree with this. So the Arc Angels are sentinels of Great Light, resonance and power with unique principles and personalities. This is a lot for the life of a human individual living in the lower denser Earth energies. Some of one angel's qualities could cause some form of conflict with another angel's qualities without meaning to, for us smaller beings.

It's just a lot of energy for a human depending on who you are. Some people can handle it. For so many of us we always tend to think more is better. That is the tendency for human consumption. 'More is better', 'we want it all and we want it now' or 'we want to go all the way right now.' So just be careful on calling too many of the angels together at one time. Again if you are already doing this and you are not having any challenges with the different energies then you are okay to continue.

You may want to work with them one at a time or ask in meditation which Arc Angel(s) are right for you to work with on your life path. They may present themselves to you naturally in a dream, vision or in meditation and give you a message or say hello, letting you know that they are available to work with you. When you are working with the Arc Angels you have powerful allies and you will not need to worry or be concerned with dark forces, entities or demons bothering you. None of the dark beings will come around you when you are working with the Arc Angels.

Many of us have spent time in the Angelic Realm with the angels prior to coming to Earth. I have met many of you in sessions. Sometimes, you may have a feeling that something is different about you and not know why. Often this is due to your energetic remembrance of the other side. Either being in Heaven, being in the Akashic Records or being with the Arc Angels.

Angels, Masters and Occurrences

Once upon a time years ago, while I was working as a professional artist before embarking on my transition into spiritual studying and teaching, I was participating in an art show opening at a local cinema and acting school in Phoenix. The curator of the show struck up a conversation with me, sharing with me details of a very unique story that I didn't expect. He said to me "I remember you from the other side. You were there with Babaji." Now, this was someone who had never met me before and who had never studied or knew who Babaji was. Years prior, he had been driving a produce delivery truck up 1-17 north en route to Flagstaff from Phoenix, when a terrible accident occurred. This was long before they had expanded this two lane highway. 1-17 had been noted as one of the most dangerous highways in America because of it's steep elevations and hairpin turns. The truck was over turned pummeling down the cliffs.

Immediately he was on the other side and Arc Angel Metatron was there along with Jesus sitting on a large rock. He expressed to me "I will never forget the sound of Metatron's wings. It was like a clacking as if they were metallic." Jesus spoke to him about love and healing. The next thing he knew he was waking up weeks later in an ICU unit at a Phoenix hospital. The doctors had told him that nearly every bone in his body had been broken and his body disfigured as they struggled to put him back together. This man showed no signs of immobility in any way. "And I remember you being there with Babaji" This very inspirational in the moment and intriguing for me, because I had zero recollection of being involved in any way.

This is before I had learned that I was working with individuals on the other side unbeknownst to me. One woman contacted me in California for a healing session. She told me during the session that I had appeared to her in a dream and told her my name. She immediately called a local spiritual store and found out how to get in touch with me. Several individuals over the years have mentioned that I would show up in their dreams and teach them. I had no remembrance of any of these, though I knew to trust the dynamic, because years before I had a friend that journeyed during his sleep to work on people. Another person had went into a library during a dream and pulled my book off the shelf, read the name and then upon waking went online to order it. I soon after worked with her in sessions.

Babaji's presence in my life has come through several times. When I first walked into the Self Realization Fellowship Temple bookstore in Phoenix after attending my first Sunday morning service there, I looked at a book on display and open to a page with an image of the immortal Indian saint, Babaji. I had the feeling 'I have to know who this is!' Not long after that in my first Akashic reading, my first teacher's words out of his mouth were "Babaji was your sat guru in a past life."

After that time while still working as an artist, I had painted a 6 feet x 6 feet painting of Babaji which was in my studio. I had also painted him in a mural along with many other Ascended Masters which I'd taken and displayed at the Wesak festival in Mt. Shasta, CA. One year I wasn't able to make it to the show and I was crushed, feeling like I had failed my spiritual journey.

Sai Baba appeared to me in a dream that night. He was an Indian master who was, at that time, still in human form. Sai Baba spoke with me in the dream while all the walls in the studio moved. Alive, the walls were a wrap around moving energy painting and were full of moving strokes of paint colors. I immediately identified this as Sai Baba's work. I was so excited in this lucid dream, that I took time to go tell the other resident artists that Sai Baba's art was in my studio. As Sai Baba sat with me and talked with me, he then after a while transformed into the immortal Babaji. He was cradling me with his arm around me and gently rocking me. When I awoke from the dream I was so excited, that the discouragement of not making the big Shasta show that year had vanished. They were letting me know that they were right here where I was.

Sai Baba another spiritual confidant for me, like Babaji, has appeared on the other side to many people I've worked with in session, whether they knew him or not. Sometimes my clients during the intuitive Akashic readings would ask 'Who is the Indian man with big hair wearing an orange robe?' Smiling I would tell them. "That's Sai Baba."

On another occasion in Austin, I awoke one night in a dream in the apartment room where I was. The room was dark in the dream. Then all of a sudden lightning crashed silhouetting the outline of a man with long dark hair wearing Native American head feather and clothing. As the lightning crashed, I saw a flash image that my feet were placed on a large drum and that the man was striking the drum with a mallet. It startled me in the lucid dream. Again the lightning crashed as the drum was struck illuminating his figure in the dark, and once more a total of 3 lightning crashes and 3 thunderous drum strikes. On the 3rd strike his voice said "There is darkness in this room," as he stretched out his arm and pointed his finger. A light shot out across the room from his hand. I immediately knew that it was Babaji and he was clearing some form of dark energy that I was not aware of.

Once while in Prescott, Arizona I met a friend at a

local natural market in the eatery. She had made me a special spiritual necklace as a gift made of blood stone which I still have. She had with her a special needs boy she was taking care of, who was the son of a friend of hers. She told me the boy was gifted in seeing and sometimes messages. As we continued our conversation about our day, the young boy started to shout "Baji! Baji!" I looked at him and said "Babaji." He immediately exclaimed "Yes he's standing right there!" pointing to the space beside me.

While travelling across country in my RV I found myself back in the Sedona area for a gathering I was teaching. I went to sleep on that hot desert summer's night in the RV park. In the middle of the night I was again lucid dreaming as I felt a huge surge of golden energy that I could see in my dream vision start to surround my body. It was glowing and it had a humming frequency that I could hear in my mind. All of a sudden over my head and off the right I saw clouds, as a torso of a very handsome blonde man leaned into the room, from out of the clouds. He was wearing a light blue shirt with a golden A emblem on his right upper chest. He spoke " I told you before, not this one." The golden sphere and humming immediately stopped. I knew that it was Arc Angel Michael who responded to the incident. In my next days meditation to recall everything, I had realized that this was some sort of alien group that was attempting to borrow me in the dream state, for whatever reasons. I've had so many of these dream experiences with the Masters, the Angels and with benevolent extraterrestrials that I could go on and on. I'm sharing these here in reference to the great assistance that is available from the other side.

As I awoke one quiet early morning in that same RV landing, I perceived in the dream state just prior to awakening a huge glowing soft white glowing figure standing over the RV about 15-18 feet tall. He was wearing a hooded robe and holding a very tall wooden staff. So the presence of the masters and angels has been with me and around me in my work for quite some time. Strangely enough it has almost become second nature when they appear, but it is always so uniquely inspiring to me. A guide post letting me know that I'm on the right path.

Chapter 14

Akashic Influence on the Civilization of Man

One of the oldest questions in the world has been "Are we being influenced by an unseen force?" The search for an answer to this has inspired many seers, mystics, prophets, religions, spiritual schools, gurus and students since ancient times. We now have a better and hopefully clearer understanding of what's actually going on through multiple dimensions and layers of cosmic reality. We know of God, though truly beyond our understanding, we continue to contact Him/Her/the Universe to seek a fuller understanding through our connection. Since ancient times we imagined, what could be out there? All different cultures have their own versions of Creator and other surrounding masters, angelics and deities. Some deities were even created to symbolize the forces of nature and different aspects of reality. The Angels became known to man. As did the Masters who achieved great spiritual attributes and told of their inter-dimensional perception and journeys became great examples and living proof that there was more to the Universe than just our physical plane of existence. This curiosity has kept humanity creatively evolving spiritually throughout the ages. It also inspired the ancient practice of psychic reading, fortune telling and scrying (looking into a mirror, water or other surfaces as a gateway to see visions) among other practices.

We as humans because of our inherent intuitive abilities have an uncanny sense that there is more to the Universe than just us. We have an inner knowing that we are not alone. Through our energy capabilities of the human mind, our chakra system and our DNA, we are now aware, or at least some of us, that we were designed with the evolutionary capabilities of ascension. The religious institutions have used the promise of everlasting life to curb, guide and regulate our appetite for evolution, ascension and expansion. In an overview the organized religions were created as an attempt to do just that - to organize and regulate spiritual expansion.

One of our creative tendencies is organization. So from that perspective it was never a bad thing. After the idea of organization came the creative idea and need to regulate and control. The larger any organization becomes in any sector of society, there arises the need for control and regulation. Mostly because, while we have this great possibility of evolving into our full potential, humans by nature can be unruly, over emotional and unpredictable. This is what we call the unconscious primal self. It is ruled by ego and group ego. More than one master through history has commented that 'Everyone thinks that they are doing the right thing.'

But what if I told you of the ancient story of a Great Temple in the sky? A library. The Library of Souls that we return to after each Earth life. Where we have a Book of Life which contains each Soul's planetary visits to the Earth plane. It is hidden from us just like our memories of coming here many times before. In early times people started seeing this place in dreams and visions. It became known to adepts, spiritual teachers and mystics as the Akasha or sky library.

As a special place in the heavens, a way station for all Souls, it was, like so many other higher truths, protected or hidden from the masses. It was guarded by mystics who had codes of ethics throughout time, careful about letting it fall into the hands of the wrong people, groups or the church. The new church already new of it's existence, but disregarded it as to having any importance while deeming their own hierarchy as the moderators for humanity. Then again, a lot people are not interested in something of such high order unless they think that something is being kept from them. This is our primal nature of curiosity and disinterest.

For these reasons the awareness of the Akashic Records has been hidden from society increasingly through the ages and forgotten through time. Now here in this time period where knowledge and instant communication is now commonplace, it is time for the re-emergence of the knowledge and wisdom of the Akashic Records to be more openly acknowledged and shared so that those who are ready to awaken and know the True Self have access to it.

Currently so much inter-dimensional and quantum scientific/spiritual information is available to us about ourselves, the world, our history and the Universe. The Akashic Records poses no threat to institutions of the world. It should be recognized as a bridge to help all Souls and their physically incarnated selves organize and clean up their own karmas as well as group karmas of every nature. It would cause individuals and organizations to truly recognize the karmic checks and balances, and even imbalances that keep being pushed into a future column of accountability.

The Akasha causes Souls to look at all aspects of their selves, whether light, dark or indifferent. To have Observations and Realizations as a way of re-framing our existence through healing, and self acceptance of our own checkered pasts. In so doing we are curbing the need for more ongoing colorful incarnations where we might continue along the same half asleep path with the remote future possibility of awakening.

How does the Akashic Records influence your Soul Field and your life here? Through your Book of Life. We all have an unconscious awareness of it's existence through our Soul Field, our chakras and the endocrine glands in the brain and the pineal gland, which is in touch with all dimensions. The thalamus gland or 3rd eye and the pituitary gland which regulates the cerebral spinal fluid. The crysum oil which is secreted from the pineal gland into the spinal fluid during the sun, moon phases each month. All of these personal parts of you can help you to open to a more full spectrum version of life and grant easier and clearer access into the Akashic Records.

Noted here: More and more the pineal is being associated with the term 3rd eye, possibly due to popularization of both terms. In ancient schools of study the thalamus was related to the brow center, which is what was termed to be the 3rd eye energy center. Both glands are directly linked and both participate with the upper 2 chakras.

Our Souls are directly connected to our 'Book of Life' within the Akashic Records. Your Soul, with all of it's higher knowing of the Records and the past incarnations you have lived, observes this life to transcribe the thoughts, words and deeds that you live while you are here. The Great Illusion is that we are separate from Heaven, our Higher Selves and blind to our invisible Soul while being in continual silent communion with it. Akashic Records work directly connects you with your invisible Soul body. It activates the Light of the Soul. As this continues to happen the self starts to become influenced no longer only by the ego and the primal self but now by the Soul as well.

As your Soul observes and relays the processes of life here in the physical based 'dream plane of existence' via a golden thread or chord, it acts in an etheric manner much like the Vagus nerve function does inside the human body. Not in the same shape or form but in the same manor. The Vagus nerve is like a high speed router that relays information, nutrients and energy to and from the brain to all organs, body parts and even on a cellular level. It can coordinate all functions simultaneously including sympathetic/parasympathetic nervous functions, stimulation of emotional impulses and subconscious mind functions. Wrap your head around that for a minute.

So while the Vagus nerve is mult-tasking continuously from within, performing high level and seemingly miraculous functions (the lower function system), the Soul field and Soul thread into the Akashic is outside of us carrying original DNA in our fields that also is connected to DNA existing within our human body (the higher function system). And they are both 2 interdependent systems performing higher/lower conscious functions while waiting for the invitation to come in and blend more fully with each other.

Your Soul Field performs similar functions to the mind, nervous system and subconscious mind in a higher state of being. As a radiant field of the True You, it knows, sees and participates silently with everything going on in your life. It is also the gateway to the Higher Mind or Super Conscious mind of the Higher Self. The Soul also connects to your Spiritual Guides, your Angels and Arc Angels and the Ascended Masters as well as God and the Universe. The amount of these connections is metered in any given moment depending on what is needed or simultaneously being experienced on the Earth plane. Those of us who practice consciously connecting with these higher beings or practice connecting to their

Akashic Book of Life have more of an easier or even effortless connection as needed or chosen when utilized. If you're just beginning it might be overwhelming to call in too many of these beings right from the start. Use your best judgement when connecting with the others and they will help guide you.

Even if you do over expand or try to do too much, your total energy system including your ego will limit the amount of energy coming in as a natural fail safe backup system based on balance and self preservation. We've all experienced this in our own ways. Meaning the lower mundane parts of the Self will challenge the higher frequency parts of creation and Higher Self attempting to connect or enjoin more with you. As the lower mundane self becomes more aware of your higher aspects not being a threat to it's existence, it will sooner or later start to cooperate.

Move forward in your inter-dimensional connections with grace and ease. Trying to push the tide can limit the process or create tremors. Or it could create a shift in your reality that may cause conflicts if you try to call in too many of 'them' or too much energy all at once. Though with practice it can be a very comfortable experience. Our total being learns how to have a natural tolerance to more energy flow coming in as we progress in making the process of life more enjoyable. We are energy beings so to recognize, becoming aware of and using the energy is fundamental to our purpose for being here.

While your Soul field is connecting with your physical and energetic self and then up into the Records, your brain, heart and nervous system are all silently participating in this connection. In addition simultaneously, these parts of you are also connecting with the atmospheric field here on the planet's surface which contains a frequency field with every frequency level from very low to very high. Through our effortless connection to this frequency field, our brain and mind field picks up thoughts and ideas as we draw them into our mind field, then into the brain and using our thoughts, senses and emotions, we project them out into our reality.

As this happens simultaneously through the Vagus nervous system's antenna-like subconscious and emotional functions and their energy fields we can connect knowingly and unknowingly with these frequency bands. Signals and levels of consciousness such as joy, love, happiness, anger, fear, regret, pride or depression etc. As our emotional field attracts this and is brought into alignment with the frequency bands, it is instantly sent into the body's nervous system where it radiates the emotions through the sentient tissues of the body. We then consciously / unconsciously assign names to these energies as feelings. *"I'm feeling inspired" "I'm feeling lost" "I'm happy!" or "I'm afraid."*

So while our higher chakras, heart, throat, 3rd eye including the crown chakra are naturally tuned to connect with God, the guides and the Universe, If the thought and emotional frequencies that we are tapping into with the body and lower parts of the mind are more open to the higher frequencies, the mundane world and the low frequency bands of energy, we are then having a natural energy carousel between the higher and lower self. While your heart is in the middle.

This is a naturally occurring dynamic of conflicting energies between the Divine Self and the human self. While still connected to the higher frequencies we may not be fully aligned with them as the lower frequencies are more dense and tend to block or override the higher energies. This can happen all too easily as the dense lower frequencies with the help of gravity and the physical world find the physical body and can resonate or attach more easily than we would prefer.

This is the challenge for the Divine self to come in while simultaneously being connected to a denser physical form. Once the energies align inside of us with the frequency fields through thought and emotion, we usually tend to recycle thoughts to feelings and feelings back to thoughts, while being caught in that cycle continually, until we can accept them, let them go, and rise above them. *Please refer to the 'Let it Go' and 'Body Scanning' exercises for help with the recurring or trapped emotional and thought form energies.*

These lower thought forms and emotional energy patterns are paired together in the moment as the brain uses the senses to process them and then experience them through our reality. This causes us to focus on the surroundings of the mundane world including personal and impersonal objects and situations. In your awakening process, you may also become more conscious of the negative or lower frequency thoughts and emotions than you previously were. And as we try to behave differently, we can become overly conscious or sensitive to the thoughts and feelings we are experiencing, somewhat prolonging them or intensifying them. This can create challenges, conflicts or delays in your spiritual well being or self learning program. Do not give up if you hit a mental or emotional road block. All of these things are part of the over all self development program. You simply have to identify these blocks and become more fully present and aware of them in order to fully release them completely and step into your new 'Future Self.' As this process takes place the old programs will no longer be running your life.

As we are now becoming more aware of the 'problem' that we focus on. The seeming 'problem' is that thoughts and emotions are sending that signal out to the Universal Reality Mirror, which is then sending them back to us through situations and form for us to experience and reflect into and upon. So often we get trapped in the mirroring reflection process. We become angry so the reality mirror shows us anger in all it's forms around us. Then we become more angry and the reality mirror then shows it to us again with possibly more intensity. Whether consciously or unconsciously we have to break the mirroring loop.

Sometimes we have to interrupt our own thought and feeling process long enough to get out of our own way. If this process keeps us spiralling through a lesser experience of truly living, we then tend to use the ego during this phase to focus on 'my new car', 'my house', 'my feelings', 'my friends', 'my enemies' and all of 'MY stuff!' Life becomes competitive and confrontational through the ego and the primal self. We have to find ways to interrupt the looping of these lower frequency energy patterns. Focusing continually on gratitude, compassion and forgiveness can be monumentally helpful in shifting this for you.

Hence the age old spiritual statement: "I am not this mind, I am not this body." This statement was never meant to disconnect a person from physical existence completely, but more to meter a person's over identification with the physical experience of the Earth plane and the internal/ external dialogue we have through our over indulgence of thoughts, words, deeds and emotions.

Try this: ***"I am not this mind, I am not this body, I am the consciousness that flows through me."***

As your Soul's connections through the body and the Book of Life are happening, there are events that you placed into your Book of Life prior to coming down here. While still in Heaven, with the help of the Guides, you chose the best experiences or groups of experiences that you wanted to have including the Souls/people that you wanted to connect with and have experiences together with on the Earth plane. Now in this lifetime family members, lovers, spouses, friends, enemies, with You and your help write newly revised versions of past life stories that maybe You felt needed revisions in some way or could be augmented. 'What happens if we try it *this* way?' Or 'what happens if the Souls are playing different roles with each other?'

Now on the Earth plane, you wake up gradually to the new lifetime. In a new body growing up and not knowing who you truly are as a spiritual being. Then you start your spiritual remembrance program. As you start studying everything from meditation to chakras and energy healing to psychic awareness and alien races, you come across the Akashic Records. And there is that consciously unconscious aware feeling again. "What is that? Akashic what?" Then you come to find out that you can look at all your past lives which brings a new level of self awareness and knowing.

Then you find out that you can also look at present and future events coming. As you get to this level in your meditation journeys with your Inner Akashic Vision work, you may see something that seems like it would be too much for you in this life depending on what karmic free will choices you have already developed through situations in your life. You do have the ability to augment, if not change the event from happening. If you are seeing it in your Records then you are being given the opportunity to change it.

You are not allowed to go into your Book of Life and simply rewrite it, BUT, from the Earth plane with a conscious knowing and a new choice you can change the outcomes of these future events by changing your habits and changing your thoughts, words and deeds. This could also involve principles of integrity or forgiveness and gratitude. As you change your response patterns to life, your life changes. Any coming events that you were concerned about may vanish or can be greatly diminished by changing yourself, your reactions and responses to life situations. As the coming events that were placed in your Book of Life are no longer needed in the video game-like obstacle course of Earth School.

Another associated topic is that of free will. Do we have free will? Yes we do, though we may be bound by our Soul agreements and contracts to have certain experiences or meet certain other Souls. The idea of having free will is a matter of worrying about having control or the loss of control. Remember that we are also here experiencing karmic clean up from other lifetimes as a result of free will.

Your personal growth influences the 'others' through the group consciousness and the Soul group. Over the course of time on the Earth, the Akashic Records helps the brave Souls that come to the Earth plane to experience through a human incarnation to continue to grow and evolve. To evolve through the physical body and personality and to evolve as a Soul. The gift of this dynamic is two fold. Not only personal growth and evolution but also in societal growth and evolution. The Souls are influencing the growth and evolution of humanity as a Divine race. So that one day in the future times of humanity we will experience heightened senses of natural Self Realization and Enlightenment together as a race. So if you feel like your life is inferior or that you may not really be making a difference in the world, think bigger. You are making a difference!

We can then take our place along side many other highly developed human races in the Universe as well as the Pleiadians, Andromedans, the Arc Angels and the Ascended Masters. In truth we are potential Ascended Masters in training. The graduating class of Unconditional Love is relatively small in comparison to what it can actually be, though every one is a winner in some way. We are sent here to learn to experience Unconditional Love while in human form, not just in the after life when we return into the Great Field of that Love where it is effortless to feel and experience it.

There it is easy, here it takes practice. Our natural born Earth life is a holographic dream that only lasts a short while compared to the Eternal Now time of Eternity in Heaven. So when we come down, the gravity of our atmosphere gives us a different reference to linear time and space through 3rd dimensional reality. Some Souls are evolving more quickly and some more slowly meaning possibly more incarnations on Earth. While all Souls are equal, they are not on the same levels of evolution. These varying levels of evolving Souls are synchronistic and yet seemingly chaotic in their placement among each other for interdependent support and also to level each other up. Here on this planet we value life. We also value equality with our freedom. Though, all Souls are not equal in that they are developing at varying rates. Down here on the planet we are not measured by our Soul development level, only by our race, IQ level, sports competition, our achievements, our belongings and so on. We only simply tend to know that all Souls are Divine. On the Earth plane we meter this physical self, though not the spiritual self, which is also growing and evolving.

So while in a sentient world we tend to seek out comfort, all the while knowing that on a Soul level the more uncomfortable the situations we go through that there is great healing, awareness and progress being made. Sometimes quicker, sometimes slower but we are all evolving ourselves and each other. Have you ever noticed that there is that one person that stands out? That is helpful to everyone around them or that helps their friends and family to grow? And while they may feel alone or completely out of place, they are one of these more evolved Souls. Maybe that person is You.

Exercise 9

Journey into the Akashic Records for Healing & DNA Activation

Guided Journey Meditation - Always remember to never do this work while driving a vehicle or using any type of machinery. You must always be stationary in a quiet meditation. Read the script and then close your eyes and visualize the journey.

As you sit or lay down getting comfortable, really let yourself start to relax more and more. Breathing in and out very long, slow and evenly. Just let all the cares and the concerns of the day fade away. Relax every part of your body from the top of you head to the tip of your toes. Become present with each part of your body as you relax that area letting it gently release. As you continue to relax your heart rate slows down ever so gently... and now relax your mind with gentle peace. Just allow it to really slow down knowing that all is well and as it should be. Relax now your facial muscles, your forehead and your jaw. Allow your mouth, lips and tongue to relax. Allow your eyes, eyelids and your eye muscles to release and relax ever so gently. Relaxing your ears and your hearing.

As you quiet and relax down even deeper taking the time for your body, mind and spirit, visualize a beautiful golden sphere of energy around your head and chest. A large sphere that extends up over your head. This is your Soul energy. As you see it, start to breathe this energy down into the top of your head. Breathe the golden light of your Soul into your face. As you breathe the golden light in now, coming down into your chest and heart. As you continue breathing the golden light of your Soul down around your body and through your body until it fills you up completely. Fully coming into your body lifting you up gently and supporting and soothing you in every way.

See now a beautiful portal of light coming from above as it comes into the room. As you continue to relax and focus on the portal of light, the more you look into it you notice that it extends farther and farther up, higher and higher. As everything else in the room starts to become just slightly dimmer and fades back, the portal of light becomes more prominent and present around you. As it continues to grow brighter, it gently seems to be calling you into it's light. As your body starts to feel lighter and lighter now, you start to levitate and float up into this great Divine Light. Floating gently up the higher you go the more you seem to be able to view into this portal. The light extends all the way up through the heavens. As you see the stars and celestial bodies. You see a great field of white clouds just above you. They are glowing against the backdrop of space. As you move through the portal of light into the clouds, you are passing through the astral plane. You are now passing through all of the dimensions for reaching the higher plane of existence. As you come out of the clouds you are coming directly up into the Great Hall of the Akashic Records. The Hall is quite spacious with great skylights overhead in the arched ceiling. The great stained glass windows lining both sides of the Hall are beaming in the cosmic light of creation. As the great beams shine down in streams of colored light that light up the Great Hall.

You have come to the Akashic Library today for a very unique reason. You a here for rejuvenation and activation in all areas of your life by your Divine Birthright of Your and your Higher Self. As you stand in the middle of the Hall on this journey today you see luminous blue fields of light overhead moving slowly in a circular pattern. You also see your Book of Life open on a great table before you. You feel your own Soul energy resonating from the Great Book.

As a great sphere of blue light comes down out of the field and hovers in front of you over your Book of Life, it now shows you luminous patterns within the sphere. This is your original Soul DNA. As you start to feel a connection with your original DNA it feels clear, energized and calming. As this happens you see and feel your Soul field around yourself start to glow with a luminous gold. We now call forth your original Soul DNA to come forth and to connect with your Soul field. We ask your original DNA to energize your Soul field. Your Soul now glows even more, as it sings with the energy of recognizing it's own original patterns. We now ask your original DNA to charge, cleanse heal and re-energize your Soul field. Take a moment here and just be present to the great rejuvenating process that is now taking place.

As your Soul field is now recharged with it's original luminous energy and DNA, we now invite your original Soul DNA to come down inside your Soul field and to reactivate and overwrite any conflicting or limiting DNA patterns within your auric fields.

We now overwrite and clear any and all limiting self beliefs in your auric DNA.

We now overwrite any and all limiting or negative mental patterns in your auric DNA.

We now overwrite any and all limiting or negative emotional patterns in your auric DNA.

We now overwrite any and all limiting or negative karmic patterns or loops in your auric DNA.

We now overwrite any and all limiting or negative physical health patterns or loops in your auric DNA.

We now overwrite any and all limiting blocks to your abundance in all areas of life in your auric DNA.

Take a few moments here and just be present to the great clearing, rejuvenating reset process that is now taking place.

We now invite your original Soul DNA to come down inside your physical body and to reactivate and overwrite any conflicting or limiting DNA patterns within your physical being.

We now overwrite any and all limiting or negative mental patterns in the neural net and brain with your Soul's DNA.

We now overwrite any and all limiting or negative mental patterns in the nervous system and the Vagus nerve with your Soul's DNA.

We now overwrite and clear any and all limiting or negative mental and emotional patterns being stored in the physical body tissue with your Soul's DNA.

We now overwrite any and all limiting or negative patterns of lack in your body and install abundance with your Soul's DNA.

We now overwrite any and all limiting or negative ancestral patterns in your body and DNA and install healing and clearing abundance with your Soul's higher DNA.

We now overwrite any and all limiting or negative patterns of dis-ease in your body and all it's systems and install the abundance of love and healing with your Soul's DNA.

Take a few moments here and just be present to the great clearing, rejuvenating reset process that is now taking place. As your total Soul, Auric and Physical Selves are now being rejuvenated, re-energized and re-harmonized.

The blue field now moves hovering aside the Book. Out of the blue field of energy from above, a Vision Board now floats down and hovers over your Book of Life. The Vision Board is now showing you pictures, images and Visions of your Future Self. Sneak previews of the Great Future Moments that await you. Take a few moments here as you look into the vision board at your new future. You may look at where you are living, who you are with, what are you doing creatively in this future. How has your life and success changed? How has your health and physical body changed? What new things are you doing that you have always wanted to move into but the old patterns were keeping you stuck? You are free now to create and move forward with your new life.

We now bring your Future Self's attributes of abundance in all areas of life into your Present Self.
Spiritual, Mental, Physical, Emotional, Relationship and Financial Well Being from Future Self into the Present Self.

Spiritual, Mental, Physical, Emotional, Relationship and Financial Well Being from Future Self into the Present Self.

Spiritual, Mental, Physical, Emotional, Relationship and Financial Well Being from Future Self into the Present Self.

Take a few moments here and just be present to the great clearing, rejuvenating reset process that is now taking place.

As the process is now complete, the blue vision board screen now becomes transparent and fades down into your Book of Life. The Soul Sphere floats back up and returns into the field of blue light. As you now feel more complete, clear and full of love, knowing that all things that needed healing and completion are now perfect, whole and complete. And all possible and probable opportunities that are coming will be strengthened. You have so many wonderful things that are coming to be grateful for. Giving thanks here in the Great Hall of Records for the experience of working with the original DNA of your Soul, let's move back across the Hall towards a portal of Light.

Giving thanks for the opportunity to do this work, as you now turn to walk back across the Great Hall, you step back into the portal of Light. It now transports you back down through the heavens and back down to Earth. Coming back into your home and your physical body take a few moments to reintegrate back to the here and now.

We thank the Keepers of the Akashic Records for facilitating this Great Soul Awakening within ourselves, our lives and our earth bound reality. Om Peace Amen.

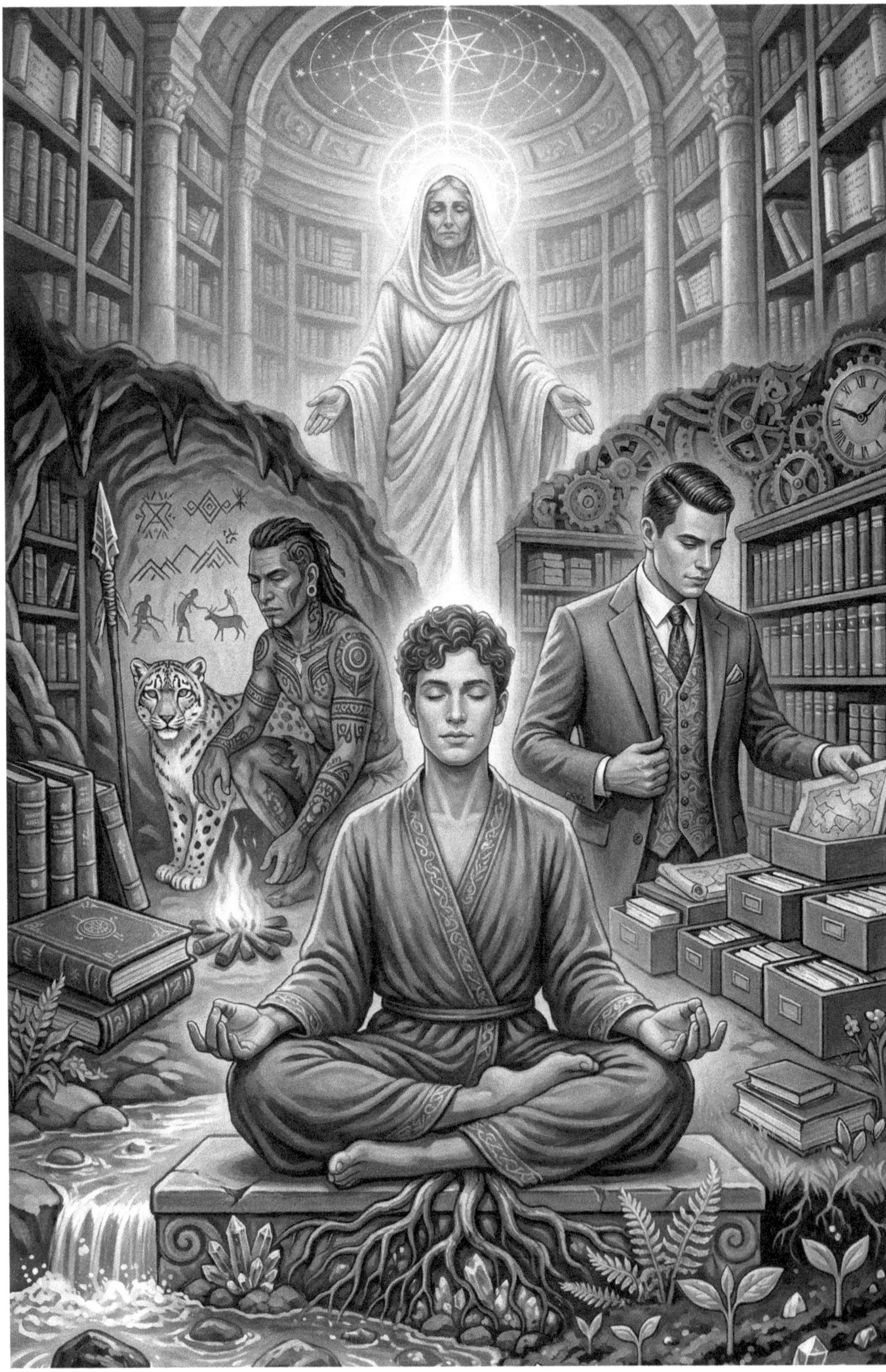

Chapter 15

DIFFERENT VERSIONS OF SELF & THE REALITIES THEY OCCUPY

Self - A person's essential being that distinguishes them from others, especially considered as the object of introspection or reflexive action. A person's particular nature or personality. The qualities that make a person individual or unique.

There are multiple levels or layers of Self. As consciousness operating through a human body, this is our primary platform of observation. 'Our' consciousness has an intellectual and a visceral experience through our energy bodies and our physical bodies. Consciousness flows in from the Soul field and radiates continually into us and out from us in spherical and spiraling patterns as well as waves and straight line directional paths. The energy of consciousness radiates into our crown chakra, then into the brain and the mind field around the body simultaneously. Then down through the chakras and the nervous system. Then out into the physical external reality through the senses of the head and body. The ancient Egyptian syllable for this movement was 'Art' or upper openings of the body. Consciousness then flows from the brain down through the nervous system with the Vagus nerve being it's primary pathway of communication into the body systems. Meanwhile the spinal chord delivers impulse communications down the spine and coordinates with the chakras (the nonphysical energy centers that float along the spine in the cerebral spinal fluid). As the consciousness is orchestrating the physical body and it's energy fields and systems, it's also radiating down through the entire body in quantum waves as it communicates with the DNA in every cell of the body through the DNA receptors at the ends of the chromosomes called telomeres.

As this whole set of simultaneous energy transfers are taking place, we have other versions of self that are in rotation. Next up are:

The 3 Faces of Conscious Perception.
God/Universe
is operating in us, as us and through us.
Higher Self/Soul/Higher Mind/Guides
are operating in us, as us and through us.
Lower Self/Lower Mind/Ego/Child Self/ Unconscious Primal Self
are operating in us, as us and through us.

And you are the lense. These levels all take turns and some of the time they are operating simultaneously. These overlays along with our life experiences in this life and past lives create our personality.

God and the Universe bring through levels of cosmic energy, unconditional love and a higher knowing that we are connected to all of creation. The Light of the Universe emanating out from the Center of Creation is Pure Cosmic Light of Creation. This cosmic energy creates stars and suns, worlds and civilizations. It radiates from the Great Central Sun of our known Universe, the home of God. God is the Collective Soul from which all Souls are born. Each Soul is moving out from the Center of Creation and starts it's journey to take on a physical presence. To seek enjoinment with life forms and start a co-evolutionary process. This is the beginning stage of ascension. As the Souls evolve through continual lifetimes in different star systems on myriads of planets.

The Great light of the Central Sun also radiates out through portals of energy. It projects a portal to and through all stars in the Universe creating a network of star gates through which the Souls travel to reach distant worlds. Most Souls here on Earth are from other civilizations among the stars sent here to continue their own Soul evolution but also to help our planet and species to evolve towards ascension. This pure Cosmic Light also is equally the pure unconditional Cosmic Love. God's Love, which is beyond our understanding (though we are always using it creatively). So perfectly and purely this Cosmic Creative Love Energy is the Light that is Love and the Love that is Light. This cosmically creative love light is used to create, evolve and build everything in all dimensions. We often think of light and dark as being opposing equals. For some studies in balance this may be helpful such as the Asian concept of Yin and Yang energy and how they work together. Though, for all practical knowing and studies here, the nature of the Universe is Good or God, because the nature of the pure Cosmic Love Light Essence is Good or God. The Good far out weighs anything less than good. And it is only a matter of time before the Good will transform or level up anything less than good. In the meantime welcome to the Cosmic Reality Show now playing in a neighborhood near you!

Higher Self / Soul / Higher Mind and the Guides bring through the next level of the lense. While channelling the God Self and Universal Energy they help to focus or funnel these energies into a more palatable form for the Earth bound self to utilize. They also bring in a higher functioning level of intellect and have access to the higher bands of frequencies available to us that show up as virtues, values, inspiration, ideas, motivation, protection, blessings, synchronicity, serendipity, manifestation and so on. This part of us is the gateway for higher levels of consciousness processing including self study, love, self realization and enlightenment. The Soul processes observation, past life and alternate reality options for enhanced levels of growth to be processed spiritually, energetically and experientially on the Earth plane.

Meanwhile the Lower Self / Lower Mind / Ego / Child

Self/Unconscious Primal Self, while also being aware and involved with all of the above, processes more immediate interactions on the playing field of life's ongoing reality show. Since we are most in touch with our physical surroundings and all of the other people in it including friends and family, The Ego has a deep visceral grip on our life functions mostly out of survival. It is in constant contact with the Unconscious Primal Self and together they can control a large part of our reactions and responses to life situations and even more so to any perceived life challenges.

The Child Self has all of this: Life's early influences, remembrances and learned responses, and is also easily influenced by the Ego who assures it that it has it's best intentions and well being in mind. The Ego also influences, not only the Inner Child but also has the ear of the Subconscious Mind. When the Ego and the Unconscious Primal Self are in control of the Subconscious we can get a colorful variation of new life events being presented to us in and along a recurring theme of past traumas, negative situations, similar repeating relationship challenges, etc.

So, the Lower Self as the complete physical body and all of it's energy systems can either be under the influence of the Child Self, the Ego, the Unconscious Primal Self and the Lower mind who has access to the lower varied frequency fields in our atmosphere that generate depression, fear, apathy, anger, hate, jealously, rage, terror, hopelessness, etc. Through all of this we can find ourselves looping through old patterns which can be laborious to our Souls's evolution and our Life Path. But the Lower Self also has access to the God/Universal lense and the Higher Self, Soul, Higher Mind and Spiritual Guides. So we are in the middle of all three of these lense levels as they shift from one to the other and sometimes overlay each other. This makes for an interesting and colorful daily journey. Most people never wake up to this and here you are reading about it while you are experiencing just that.

So what is the answer? Balance. Meditation & Prayer, Spiritual Study and Self Study. Healing, Forgiveness and Gratitude. Healing the body through better diet and incremental cleansing methods, such as colon cleansing, liver and kidney cleanses, and even ionic foot baths can be very helpful. All of these things can bring balance in such a way that the negative functions or masks of Self can be laid to rest. Such as the Unconscious Primal Self or the 'animal within us'. The Ego, can take a back seat to the Soul and the Higher Mind bringing much less inner conflict through our inner and outer experiences. The Child Self can be nurtured by the higher levels of Self as you learn to become your own parent. Leading the Inner Child away from the looping remembrances of childhood fears, pain or anxieties. This can bring less confusion to a person's life as the Ego loses it's grip of submitting the Inner Child to the unconscious primal self and the Ego's fight or flight knee jerk reactions. These two partners in crime, the Ego and the Unconscious Primal Self both loosen their hold on influencing the Subconscious Mind, nervous system and physical body.

Life will start to become easier and more rhythmically

flowing with energy. The 'ups and downs' of life will not be so overwhelming as you start to realize that even if things do come up they will be short lived and not so overwhelming. This can be extremely helpful especially if you are studying energy, the Universe and manifesting. Bringing awareness to the higher dynamics of the way things actually work and occur in the Universe while still repeating the old patterns in life can be like leading a dual existence and a mixed bag of life's experiences. So take action and heal the Past Self as you move into the Future Self. The book: 'the Healing Power of the Akashic Records' has invaluable information for reprogramming the Subconscious Mind, releasing deep traumas and reactivating the nervous system, energy bodies and Total Self.

The next level to discuss are the past lives that you've lived. These are other versions of Self. These versions of you occupy other times and places in the great time space of realities. We have played many roles in many lifetimes on many worlds. As we incarnate each time, we take on new goals and attributes of self that we choose to experience in our Soul's evolution. While the Soul is growing, observing, learning and experiencing, the enjoined physical persona is also growing, learning, observing and experiencing through thought, word and deed. The Soul has the ability to help the physical incarnation naturally evolve spiritually and creatively through the body's energy systems, nervous system, chakras and mind. As the individual may find their balance for the process to continue, the individual then we may have the opportunity to step towards natural levels of ascension.

The other lifetimes we have lived can have positive and/or perceived negative influences on a person's current incarnation. This happens through the stored energy in the DNA of our bodies, energy fields, our Soul field and the DNA stored in our Book of Life in the Akashic Records. Our Soul Body while connected to the physical body on the Earth plane is also connected up into your personal Book of Lifetimes or Soul Records. The greatest realization in our journey be known, is that you are truly a Miracle of the Divine just by being part of this amazing dynamic. Once you see the world, Heaven, the Universe, the Akashic Records and your Soul from this perspective, it can become challenging to see all of the many varying situations in this world that create such astonishing levels of turmoil. It can also bring harmony to your perception of the seeming conflicts of the outer reality in that you start to recognize that everyone else is playing out their needed and agreed upon life projections both individually and in group consciousness. It can be a lot to take in and at that point focus on staying balanced on your journey.

In relation to the past lives, time operates much differently than we think it does. Time is more like a ball of yarn with continually overlapping points. On Earth we have created linear time to give us a simplification of the Now Moment, what just happened and what will happen next. We have learned to live our lives by past, present and future. We have conditioned ourselves to live a certain relative life span. Time truly is relative. So in the

Mind of God or Universe, the Eternal Now is occurring. In the vastness of Universal Time our past, present and future lives are all happening simultaneously along side of each other. Your Soul has reference to all of the other lives being lived, though, through your consciousness your primary point of focus is in the 'Here and Now' in this life.

This greater dynamic of Universal 'Here and Now' is the reason why that when we do emotional childhood trauma release and healing, it not only heals the past, it can heal us in the present moment and our future. So what can we do with this idea that all lifetimes are being lived now and that some are positive and some may be negative? Well your Soul's focus is on this lifetime, and while I've studied this dynamic for years, I've seen negative dynamics occur in a person's life that couldn't be explained by all of the positive spiritual work they had been doing. Nor, could it be explained by astrological limitations or processions.

The astrological influences may be in tandem with the events or they may be unexplainably not in tandem. The understanding came from messages by looking into the Akashic Records and seeing influences from other lifetimes being accessed through doorways that the Soul also has access to. As this was seen, I could turn the Soul's focus away from a negative lifetime's doorway and literally create an new and recently unforeseen level of healing and clearing in a person's life. This is not something I would recommend dabbling with, but I'm using this to illustrate the dynamics of the Soul. If you feel that you are in need of assistance on this level you should reach out to myself or a good shaman. Sessions are available at my site.

Potential Realities are another part of the whole Universal that can affect us and our current reality in a myriad of ways. The primary way to approach this is from the perspective of Past Self and Future Self. And the positive influences that we would like to have, instead of any negative influences that we've experienced from the perspective of Past Self or possible negative variations that could occur to the Present Self. From this primary outlook we have a whole lot of colorations that can be anywhere between the ideas of good and bad or light and dark influences that we contextualize from the myriad of situational responses that occur in each of our lives, through us, as us and to us.

My reason to designate the happenings as good or bad is not to bring challenge to your ego or perceptions but only to simplify the idea for you. In the 'in between' space of such pointed events along the event horizon we have a whole lot of indifferent responses that could be perceived in either direction of designating between positive or negative. And...there is usually more than one reason behind everything happening as it occurs on multiple levels in tandem with our life lesson learning and that of others.

Here's the good news. Infinity is a concept that has already been tried and proven through quantum physics. In the infinity of inter-dimensional Universal variations there are an infinite number of possibilities. Put in another way, there are versions of you and your realities that can be accessed by your total being in this current journey. Your total being includes the integration of the different lenses of Self that we have mentioned earlier such as Higher and Lower Self, Ego, Inner Child, God, Subconscious Mind and total self.

Through these interactive lenses of Self you play out the version of primary reality that is most helpful for your learning and experiencing on the Earth plane. It can be looked at differently than just manifesting or creating a new experience from within this primary reality even though our point of reference is 'here.' We want to bring it 'here.' From the quantum space of variations of different realities, it is possible to consciously, subconsciously or even unconsciously access one of these alternate reality time lines and move it into and through your existing reality. It's always better to do this knowingly as opposed to unknowingly.

This is where you learn to focus on intention. Life can and will give you everything that you willingly intend to have, if you become convinced without conflict that it is rightfully yours. Soul's Purpose and a positive intention to help the world in some way can supercharge this. Be watchful of justifications by the ego of what and why it wants something. It's always better to get the ego on board with a higher purpose than to bend to the will of pure ego. Having things just to have them is pointless and does not serve the greater good. So you may be able to create on that level so that are shown it can be done, though at some point your Higher Self or various guides and teachers can intervene and shut it down, so that your ego doesn't manifest calamities in your life. The kind we hear about in the world everyday.

When you are now aware of this process it can make it easier for you to control or oversee what happens for you interactively with your reality. When you are aware of influencing your Subconscious Mind through the conscious mind prompts you can have a more guided and directed experience. When your Subconscious or even your Unconscious Primal Self (sometimes with the guidance of the Ego) pulls at alternate realities, you may get a mixed bag of experiences that could show up as challenges, lessons or even unwanted road blocks to a healthy and happy life.

Not to worry, your Higher Self, God and your Guides can intervene in such situations that may start to occur and put everything back on track for you. So in that, it could seem like just more of a detour to your desired path instead of a complete road block. There is always a way no matter what happens in life because you are a Soul in human form, sent here to love and create and to experience all of the colorations that make life on the Earth plane so interesting.

The old saying 'Ignorance is Bliss' can hold some comfortable meaning for evolving Souls who are not ready to look at the realities and the Universe on a deeper study level. And yet, for them everything will occur perfectly within their scope. There are those of us who choose to look deeper into the realities and to create within this reality as well as choose, create or manifest a new variation from reality variants within the multiverse. As long as we are operating with reverence and virtue our efforts will be acknowledged for the better. Time and timing are other components that interacts with such requests. Some things can occur straight away while other things can take longer.

Let's look at manifesting from this known reality. The Universe is like a great mirror. When we mirror anything to the Universe, it is mirrored back to us in our reality. Some things quicker and some more slowly. So the Great Mirror has infinite distances of time variations for the mirroring effect to occur. In this, is the example of things that we have said, visualized or asked for and may have been long forgotten because they did not show up right away.

One of the things that most confuses us as physical beings in a mirror Universe is that we send something out through the reflection process and then when the reflection occurs we become directed or distracted by it's reflection and may even think we are being trapped by this reflection as an event manifests. As we move, our reflection in the mirror moves, then we try to outwardly manipulate the reflection, instead of changing and redirecting our own movements. We become outwardly directed by our own reflection we are sending out.

It is most helpful in these moments to remember 'As within So Without'. In this way, we can continue to guide and direct the reflection more effectively and smoothly in our reality. This process, when practiced continually, will bring a greater sense of self awareness and responsibility. The mirror is always watching or reflecting back to us. This is a whole new perspective of the Universe and reality as the interactive all pervading Universal God Source. You are in it and it is within you. As within so without. The Holographic Universal Reality just became more interestingly and interactively deeper in this model.

The Higher Self is your bridge for success and evolution. Some of us are able to naturally focus on an abundant way of being and living because of either their previous incarnations or the support and nurturing that they have received in their childhood and growth development. These people operate 'as if' and they communicate with the world around them in naturally higher frequency and response patterns. They are not completely impervious to the world and it's throws, but are more accustomed to navigate it's challenges with more grace and ease. Some of these people even welcome challenge as a known way of growth and adventure.

For others, at times life and the world around them can be a mine field of ups and downs. Maybe because of their varied results in life they are focused on a somewhat separate belief in a Higher Power or maybe even devoid of anything sacred or Divine aspect of a Higher Power or version of God outside themselves.

Then there are those who are studying and/or are becoming awakened to a greater truth or the 'secrets of the Universe' and they may struggle with the different levels of learning, experiencing, karmic growth patterns and lessons. While studying the law of attraction, the chakras in the human body, the spiritual deities and masters, they may struggle with their own invisible fears and past self primal grip of the Ego and the Primal Unconscious Self. This is an old repeating story in history and a natural process to go through.

This a can create a roller coaster of unwanted 'ups and downs' as they struggle to step into a new way of being or releasing the old programs of Past Self. At times they may be experiencing streaming moments of connecting with a more desired version of life. This struggle of the old self and streaming of the new self through a more desirable existence can create an on going paradox between Past Self and Future Self. This can hold people at bay for indefinite periods of time and can cause feelings of failure, procrastination or even self doubt.

For many, this can keep them in a holding pattern of evolution, not only in life, but on a Soul level as well. Time is relative and while we live in an ultimately connected and loving Universe, consciousness can be indifferent or seemingly impersonal in it's connection through us and in it's streaming connection to the Universe and our external reality with it's mirror like qualities. It can be lovingly impersonal of our timing as it is waiting on us to really 'get it.' If we aren't able to take advantage of an up level at one crossroad, another will appear up ahead.

Our Soul and 'they' are waiting on us to become more self aware. The Higher Self can help us with this. When we are in throws of life's ups and downs, it seems that we are reacting to our external conditions. Our Ego is in control through survival mode and creates any number of self looping programs of self importance, greater than / less than with situations based in fear, lack, anger, jealousy, competition, etc. with others or our external reality. Meanwhile the Subconscious Mind while trying to take prompts from the conscious mind to get things accomplished in the moment could be influenced by the Ego and Unconscious Primal Self creating lesser or more 'colored' limited versions of any desired outcome. During this time, the nervous system via the Vagus Nerve could be in subtle to severe levels of 'fight or flight' and not able to relax while also sending dis-ease communications out to the rest of the physical body and even energy bodies.

Connect with your Higher Self. You may acknowledge this as the Higher Self, Angelic Self, Higher Mind, Super Conscious Mind. Whatever version works best for you. It is here that we can influence the Ego to either stand down and be quiet or elevate up into the Higher Mind (connected with the Higher Self). You may talk to the Ego and communicate to it that your Higher Self has it's best interests in mind. It is unconscious of it's own primal nature of trying to convincingly instil control over a person's life while not able to know or reason that there is a better way. You cannot get rid of the Ego or quarantine it to a locked basement. This can only result in more Ego problems as it tries to break free. But you can influence it to work with you in a better concentrated effort with the rest of your team. Here is a script to help you with this:

Promotion for the Ego with a Full Benefits Package

(Read aloud with a convincing vigor while placing one hand on your heart or stomach as you feel the energy of your Ego moving from your midsection up into your Higher Mind.

"Dear Ego, Thank You for all of your service to me. You've always looked out for me. You've helped me to get through life so far. But there is a better way. All of the things that you've wanted are available to us in a better version. And I am now giving you a promotion! I am now inviting you to move up from behind the 'wizard of oz' curtain and take the elevator up into the beautiful high rise penthouse suite of the Higher Mind. You'll have wrap around windows (the eyes) and access to all of the senses and Higher Mind functions.

Here you'll be working directly with the Higher Mind and the Higher Self. You'll also be getting an energy upgrade with this promotion. You're only required to be at all board meetings with the Higher Mind and Higher Self. As far as your old patterns of fear, anger, comparison, competition, jealousy, self doubt, unworthiness, etc. , you won't need any of these lower impulses of behavior anymore. We want you to enjoy all of the best that life has to offer! So in regards to your old work functions, we're placing you on permanent vacation. You are only required to be at all board meetings with the Higher Mind and Higher Self. Here you will have direct access to growth, learning, manifesting, and all of the higher attributes, functions and emotions. You will see how much easier, loving and enjoyable life can be. Thank You and Enjoy" (*Repeat as needed)*

Chapter 16

Interactive Healing and Manifesting Through Your Book of Life

When you start visiting the Akashic Records it will be most productive for you to visit your Book of Life first. In the classes and workshops this is what we focus on connecting you in with. Also, when I'm reading for others I'm reading from their Book of Life with their permission. If you're going into the Akashic Records and just looking around, it's sort of like going into a major department store without a shopping list and wandering around, then leaving the store with nothing because you forgot why you went there in the first place. I highly recommend upon entering the Records moving to work with your Book of Life unless you stop off to talk to the Guides or Angels there. After working with your own Book of Life for a while, then if you want to look at other subjects like world history or specific subjects like the arts, science, healing, etc. you will have more of a working model from access experience through your Soul energy and Akashic Soul Vision. Your Akashic Soul Vision utilizes multiple functions when visiting the Records. When most people think about psychic or inner vision work, they think of the 3rd eye brow center chakra or the crown chakra. When accessing the Records we are using our 3rd eye as well as the physical eyes, the brain and mind field, the crown chakra and the Soul field all simultaneously. Primarily we are utilizing the psychic capabilities of the mind and pairing it with the intuitive abilities and the energy of the Soul as we project into the Akasha (Sky Library).

Actually you are always connected into the Records. This is how you input your thoughts, words and deeds into your Book of Life. This silent invisible strand becomes more active and energetically strengthened as you start to communicate more with your Soul Records. As you set the intention to project up into your Book of Life within the Akashic Records and begin to meditate deeply on your guided journey process, something very interesting happens. Your Soul field that has been with you all along starts to wake up, so to speak, and become more interactive with your total self. As your Soul field becomes more activated, it's energy (Your True Self) becomes more active and energized and starts to have an influence on the physical body, it's energy systems and your energy fields.

All of the energy systems in your Total Self are all working simultaneously and for the same goal. It is for you to create, experience thoughts and emotions, and to evolve through participatory situations on the Earth plane. So your Brain and mind field are working with your senses and your nervous system. Along with your chakras and the multi-layered chakra field surrounding your body. Your cells and your DNA are also talkative and working within you. Meanwhile your body tissue can experience, in the moment, the effects of emotions and/or stored emotions via the nerve impulses and the thought and emotions being generated through the mind and emotional fields.

These mental & emotional response signals are being generated in harmony with frequency bands of energy in the atmosphere which they align with, as the self places a label on the thought/feeling response. Some say thoughts happen first, and then the feelings. Some say feelings happen and then trigger thoughts. Really it's all happening at once, or that it happens so quickly that it seems to be happening at once. When we are learning or even practicing spiritually it's quite regular that individuals, their teachers or practitioners focus on one simple thing, such as the chakras or the heart chakra while thinking that this will heal everything, it's only a start or the tip of the iceberg. Let's say it's a good place to start. And generally speaking it will heal everything, but a deeper study will yield deeper results.

It is good to have a focus for balance and a way in while all of these other functions are going on simultaneously and systematically. While someone is focused only on opening their 3rd eye, there are multiple other psychic/ intuitive components operating within to consider.

When you are looking into the past, present or future using your psychic mind, you are usually using your Psychic Stream of Awareness which moves into and through the astral plane or 5th dimension to your target of time, object, situation or person. If you are connecting with another individual whether remotely or in person you are usually using the 3rd, 4th and 5th dimensions to do so. Here's the challenge: The 3rd and 5th dimensions are very busy places with a whole host of different energies and beings coming and going. 4th dimension is different it is the mind's place of housing the considerations of time/ space perspectives. Now there are some really gifted psychics out there and some of them operate beyond these metaphysical constructs. They may be working with an E.T. or special Angelic guides that grants them access to information and they have a true to form way of conveying their messages.

You may ask "What if this psychic person is me? I could be that person!" You might be! Be conscious of the underlying dynamic that there are a lot of egos in the world looking for a spiritual 'hack.' (A way to gain quick access to info without doing the work in a cleaner way). There are also a lot of very gifted psychics in the world with big egos. If you had access to any information that you wanted about yourself or others, would you be confident and maybe even a little overly confident? Just saying, it happens. The same over confidence can keep a great psychic from gaining access to the true Akashic Records. The key to accessing the true Records is Humility and Reverence and the Real Reason for gaining access is to Know the Self (Know Thyself). You don't have to be a saint or a nun but you also might not want to have an entourage of dark or questionable energies following you around feeding you psychic messages all day and night.

The truth is that there is a reason why we are all not operating as psychics even though it's a natural gift we have access to. When we move past the ego we will have access to so much more. Do you think that the folks upstairs might have something to do with that? Could you imagine if everyone was a naturally born psychic (even though we potentially are) running around in the world creating whatever their egos desired through psychic projections and awareness? Then it truly would be a circus. Maybe it was deemed that it would be better for specifically those who are prone to go in the direction of psychic to be granted access to ESP. There are many who are naturally gifted with no explanation of it. And there are many more who are gifted and never say anything about it.

So lower level (3-5 D) psychic activity can equal lower or varied results. Try 10-12 D. This is the home of the True Akashic Records. Why is it so remote or higher up? Because this is the way of protecting the information of the Souls and the way station of the Records for connecting flights back to Earth into reincarnation. This is where the high levels of information are stored. If anyone could have access to all of your Soul level information in the astral plane, things could get very risky or choppy here. This is because the astral plane has a lot of demonic activity, angelic activity and everything in between. Wandering

Souls, entities, elves, alien creatures both good and bad visit or live in the astral. So the astral has everything and everyone that comes through to the Earth plane. Everyone and everything that comes and goes. The astral plane is a very busy place. The astral plane is a very large place and stretches out like the Universe. It behaves in this way because it's properties are energetically physical in nature somewhat like the 3rd dimension.

Healing occurs in a multitude of ways from accessing the Records. I have included a special chapter on 'Psychic Healing' which is in my books 'Journey to the Akashic Records' and 'The Healing Power of the Akashic Records' which I invite you to explore and read. Let's talk about the Healing properties that can occur as you are practicing at accessing your Soul Records. When people find out that they're not going to learn the 'secrets of the universe' right off the bat, the ego might become disinterested and find ways to keep you distracted from your practice and studies with the Akashic.

Accessing your Soul Records holds the key to the mystery of the miracle that is You. However, this was the case in ancient and modern day India as well. Students will come to the masters and ask to learn the secrets to their special powers. They want the powers for themselves. The teachers have a natural selection process that usually weeds out a great number of power hungry dabblers. Now if you are new or just starting out in spirituality, please do not be shadowed by the idea that you aren't enough or doing enough. You are here and that is great.

The Keepers of the Akashic Records know that if you

are coming to the Great Hall while you're supposed to be down here in a lifetime playing along, that if you're there to request access to your personal Book of Lifetimes, they know that you are on the right path and that you are requesting to know more about yourself.

Self knowledge can bring wisdom. Wisdom can lead to karmic clearing, growth and evolution. To Know the Self. And then you are perfectly moving forward to new improved life pathways in harmony with your Soul's purpose. If you don't know your Soul's purpose that is exactly the reason to be going into your Book of Life and asking that it be shown to you. Our Soul's purpose is so close to us and within us that it's hard sometimes for us to identify. They may give it to you directly or they may show you previous incarnations and different things that you've done during those lifetimes to see if you can make the connection. This helps you identify and to build the story of your Soul's purpose. You come back every time with another piece of inner knowing that helps you develop here during this life.

Knowing your Soul's purpose is to better know yourself through wisdom. Your Soul and your Light Body are also being ignited or charged with Divine energy. As you continue your inquiries, your Soul is activated with your physical body and you can even naturally start to clear and burn karmas that are helpful for new improved passages in this life. This is huge. It happens subtly and silently while you are doing the work.

One thing to be aware of is that your ego and primal self could challenge your newly found commitments to connecting with your Higher Self. Don't be alarmed, we all go through this on our evolutionary path to Self Knowing. The greatest expression of You is waiting to recharge and energize all parts of you. Helping you to let go of the Past Self and it's grip through the unconscious Primal Self's tendencies and the Ego's clandestine subterfuge. Helping to align with yourself more in the Present Moment and move towards your Future Self which is your greater destiny.

As you continue you will start to recognize how reality can change for you on the Earth plane. Different insights about situations that arise in ways that your Book of Life may show you literally, symbolically or both. The key is to stay relaxed and inspired in your inquiries. If you are pensive about what you might find or see this is most assuredly the Ego. The Ego makes a living by controlling your life through drama. Thoughts and feelings of lower levels re-circulating or looping and then creating real or imagined negative outcomes and the fear of future impending doom. This can keep you stuck in the astral getting more psychic messages than Akashic. The key is to stay relaxed and inspired in your Akashic inquiries.

Physical healing can also start to happen as you look into past lives and what you may have went through there. As you remain silent and view, remember you are in the energy of the Soul and the Soul has Healing Power. Pure Consciousness has Healing Power. It can dissolve the tendencies that may be resonating through the Soul from another lifetime and 'bleeding over' into the channel of this lifetime. Spiritual, Mental, Physical,

Emotional and Financial healing and clearing can occur. You may experience an up-levelling of anything from a slow steady pace to very noticeable increases in your life. When you've tried everything else and nothing has worked, go to the Soul. When you've tried everything thing else with little or no effect, go to the Akashic Records for real and interactively healing knowledge, wisdom and information. It can also help you to clear the energy around you and your life as the ego starts to take a back seat and listen more through the Soul along with the rest of your systems and Self.

Often times people have asked me, "Can you go in and re-write your Records?" This is almost never allowed. You might think you are doing that, however, the real way to change your Book of Life entries is to come back to the Earth plane and take immediate or massive action. Remember you're always connected and files are constantly being uploaded into the main frame of your Akashic Book of Life through your thoughts, words and deeds. As you focus on change in life it's helpful to bring back with you anything that you have seen in the Records you want to change as you start re-aligning the energies of reality. Remember the Universe is also watching and listening through the quantum mirror of reality. If you are studying manifestation or at least have even a simple understanding of it, that's all that is needed to make great changes. So you will be working from 2 different angles: Changing or influencing your reality on the Earth plane through Self help and growth methods (prayer, meditation, visualization, forgiveness and gratitude). And you will be conscious of whatever you want to influence with change as you are re-writing your Records from the 'Here and Now' on the Earth plane in this reality. Change can happen quickly or it may come more incrementally. Whatever your total Self feels more comfortable with or is able to adapt to, will then play out.

As you continue to practice using your inner Akashic Soul Vision and your Soul's intuition to visit the Akashic Records, you will find that your psychic abilities will continue to increase as well. What you will eventually develop is a combination of the Soul's intuitive power integrated with your mind's psychic ability and it can develop into a direct knowing or a direct seeing. Be well and enjoy the gifts of Intuition, Interactive Healing and Manifesting through the Akashic Records!

Past Life
journeys

Exercise 10

ACTIVATING YOUR AKASHIC SOUL VISION

Guided Journey Meditation - Always remember to never do this work while driving a vehicle or using any type of machinery. You must always be stationary in a quiet meditation. Read the script and then close your eyes and visualize the journey.

Before you start your journey, write down 4 questions that you would like to receive answers to. These can be anything, but should be pertinent to your life's journey. You can ask about past life influences around any certain inquiry. As you are writing your questions, be thoughtful in your asking. If you are asking a direct question around the outcome of a certain life situation, ask in a way that gives you a deeper insight as to why you are experiencing any certain situation. Don't over think it, just be mindful. Once you are in the Records and receiving messages you can always stay with a certain vision and ask more questions. Be mindful of asking why. Receiving the deeper meaning around any situation is the real reason why you are experiencing it, or if it is a future event, then what the deeper meaning would be of experiencing it. And remember you can always go back in, so don't try too hard. Just remain open and let it flow.

Now, as you sit or lay down getting comfortable, really let yourself start to relax more and more. Breathing in and out very long, slow and evenly. Just let all the cares and the concerns of the day fade away. Relax every part of your body from the top of you head to the tip of your toes. Become present with each part of your body as you relax that area letting it gently release.

As you continue to relax your heart rate slows down ever so gently... and now relax your mind with gentle peace. Just allow it to really slow down knowing that all is well and as it should be. Relax now your facial muscles, your forehead and your jaw. Allow your mouth, lips and tongue to relax. Allow your eyes, eyelids and your eye muscles to release and relax ever so gently. Relaxing your ears and your hearing.

As you quiet and relax down even deeper taking the time for your body, mind and spirit, visualize a beautiful golden sphere of energy around your head and chest. A large sphere that extends up over your head. This is your Soul energy. As you see it, start to breathe this energy down into the top of your head. Breathe the golden light of your Soul into your face. As you breathe the golden light in now, coming down into your chest and heart.

As you continue breathing the golden light of your Soul down around your body and through your body until it fills you up completely. Fully coming into your body lifting you up gently and supporting and soothing you in every way.

See now a beautiful portal of light coming from above as it comes into the room. As you continue to relax and focus on the portal of light, the more you look into it you notice that it extends farther and farther up, higher and higher. As everything else in the room starts to become just slightly dimmer and fades back, the portal of light becomes more prominent and present around you. As it continues to grow brighter, it gently seems to be calling you into it's light. As your body starts to feel lighter and lighter now, you start to levitate and float up into this great Divine Light. Floating gently up the higher you go the more you seem to be able to view into this portal.

The light extends all the way up through the heavens. As you see the stars and celestial bodies. You see a great field of white clouds just above you. They are glowing against the backdrop of space. As you move through the portal of light into the clouds, you are passing through the astral plane. You are now passing through all of the dimensions for reaching the higher plane of existence. As you come out of the clouds you are coming directly up into the Great Hall of the Akashic Records. The Hall is quite spacious with great skylights overhead in the arched ceiling. The great stained glass windows lining both sides of the Hall are beaming in the cosmic light of creation. As the great beams shine down in streams of colored light that light up the Great Hall.

As you walk across the spacious Hall, you notice how tall the ceiling is. It really allows your Soul to expand and to feel at home. This place holds the energy of your Soul and you resonate with the energy here. As you walk deeper into the Hall, you notice that there is a very beautiful table or alter at the far wall. As you walk up to it, you notice the very large book on the table. This is your Book of Life. This is where you will receive answers to your questions that you have brought with you.

You also notice a very large viewing screen that is on the wall behind the table. As you look into the Great Book, it is open to the middle pages. This is your current time in life. The Book is alive with your Soul energy. It responds to you interactively. As you view into the pages connecting with the energy. Notice now the writings and the symbols. Notice now the pictures and the images. This Great Book, your Book, is ready to interface with you providing messages to You.

As you ask your first question, relax and connect into the Great Book in a way that allows you to open and let the messages unfold, as you focus here for a few moments.

Now raise your vision up to the Akashic Vision Screen and allow your Soul Vision to activate. As your consciousness projects into the screen, the images and visions start to appear. Watch the vision in relation to your question and allow the images to expand.

Take a few moments here to receive...

As you allow the pictures, images and visions to appear

on the screen of your mind, what is your Soul Vision showing you?

When you are ready you can move to your next question. As you ask your second question, relax and again connect into the Great Book in a way that allows you to open and let the messages unfold, as you focus here for a few moments.

Now raise your vision up to the Akashic Vision Screen again and allow your Soul Vision to show you images or a movie. As your consciousness projects into the screen, the images and visions start to appear on the screen of your mind. Watch the vision in relation to your question and allow the images to unfold and expand for you.
Take a few moments here to receive...

When you are ready you can move to your third question. As you ask your third question, relax and again connect into the Great Book in a way that allows you to open and let the messages unfold, as you focus here for a few moments.

Now raise your vision up to the Akashic Vision Screen again and allow your Soul Vision to show you images or a movie. As your consciousness projects into the screen, the images and visions start to appear on the screen of your mind. Watch the vision in relation to your question and allow the images to unfold and expand for you.
Take a few moments here to receive...

And when you are ready you can move now to your fourth question. As you ask your fourth question, relax and again connect into the Great Book in a way that allows you to open and let the messages unfold, as you focus here for a few moments.

Now raise your vision up to the Akashic Vision Screen again and allow your Soul Vision to show you images or a movie. As your consciousness projects into the screen, the images and visions start to appear on the screen of your mind. Watch the vision in relation to your question and allow the images to unfold and expand for you. Take a few moments here to receive...

You are learning to use your Inner Akashic Soul Vision. The more you come back here into the Records Hall to connect with your Great Book of Lifetimes, the more you will expand your vision. This is a safe place and this is a protected space, so you are free you relax and expand as you really enjoy connecting with deeper levels of your Soul, your life's journey and other lifetimes that you would explore.

Giving thanks for the opportunity to do this work,As you now turn to walk back across the Great Hall, you step back into the portal of Light. It now transports you back down through the heavens and back down to Earth. Coming back into your home and your physical body take a few moments to reintegrate back to the here and now.

We thank the Keepers of the Akashic Records for facilitating this Great Soul Awakening within ourselves, our lives and our Earth bound reality. Om Peace Amen.

Chapter 17

Creative Perceptions For Understanding Change

As we move through our life path, through growth and change, we are not always aware of when the changes are taking place. Sometimes we are very aware in the moment when we have a breakthrough. These moments serve as uplifting inspirations for us along our journey to let us know that the Universe is listening and that we are never alone. And even then as we have these moments of inspirational acknowledgement we may not be aware of the real changes that are taking place on a deeper level or how it will serve us many years in our future. One of our greatest underlying abilities is being able to accept change in our lives. To welcome it or to create it.

Working with the Akashic Records is all about finding

your True Self. While you may be working with other forms of motivational training or books, the Akashic studies provide something deeper to add to your story. Mainly because it is *your* story. It provides the way forward to look into your Soul's tendencies in other times you've lived. How you've responded or reacted to things. This gives you deep insight on what you are working on currently in this life, such as love or how to be more loving. Maybe you've experienced many colorful events that have pulled at your heart strings both in a good way or at times maybe less than desirable.

As you've landed in this life with all of the colorations from previous incarnations you may seek to understand why you always make decisions around relationships the way you do. Why you may be alone or even choose to be alone. This can also be related to life lessons, previous experiences repeating themselves or brand new experiences that you are writing into your Book of Life.

As we study spiritually we may look into Buddhism or Taoism, Christianity or Paganism. Astrology or Yogic ways of life. What have we learned from these paths? The Akashic can compliment any and all of your studies by showing you why you may be so drawn to them. Or it all may be a brand new study for you, maybe through being drawn to spirituality or past life regression. One tendency with enquiring from the Records is to assume that everything we're going through may be hinged on something we have previously lived. Sometimes this is the case, though not always. This is a question to ask when you get in.

Some of you may be great beings from other star systems that came here to help humanity and the collective grow just by being here with your vibration. Every thought, feeling and action whether you perceive it as good, bad or indifferent serves the greater purpose for evolution of yourself and the collective. You may have energetic imprints or subtle to very pronounced remembrances of your greatness. This can come through your mind, your body, your chakras, your nervous system or your DNA.

It may not be easy for you to navigate the Earth plane at times, while in other moments you are able to create great things in your life. This can happen for a number of reasons. It may be due to the ancestral programs running in your DNA. Many of our ancestors while living their lives may have went through tumultuous events. Disease, mental and emotional upheavals, wars, lack, or family quarrels. Your body spirit and DNA can still have the imprint of these events influencing your current life. What's more is that you may have even lived other lives as your ancestors instilling even a greater influence from your body's DNA.

You are the next generation. You are the light (Soul) on the Tree of Life (family tree) that has awakened. Your job is to not struggle against these ancestral DNA patterns which greater activates it, but to seek healing and regeneration of your complete Self through new studies like you are doing here. So many go through this and remain unaware their whole lives.

This is a key reason for so much ego and primal self struggle and suffering in the world. They are simply unaware and not willing to do the work to re-invent themselves. Many do not know how or simply haven't found their way onto their own version of the Great Path yet. Many have had breakthroughs and risen above such adversities. We hear many of the rags to riches stories or stories of a great awakening spiritually that changed a person's life forever. This could be your story. This could be the story of great masters in ancient and even current times. For everyone we know of, there are so many more that humanity is unaware of. But the world and the Universe is aware of them in other ways. These are the great community builders, movers and shakers, leaders and servants to everyone around them.

When we start to awaken we can expect conflict from our Ego and our Primal Self telling us that the old way of living, existing and doing things is all we were meant for. Many wake up to the notions of their own greatness and are pummelled by the Primal Self into submission while their Subconscious Mind continues to loop them through new and colorful experiences along the same old theme. This is a knee jerk reaction to survival existence.

We must remain vigilant with the conflicts and make new choices of thoughts, emotions and actions. Practicing Gratitude and Forgiveness are your greatest tools in up levelling your path. Letting Go. As we change our thoughts and feelings, the body could play out the conflicts through sickness, injury or dis-ease trying to loop us back into the old primal pool of being. The body sometimes is the last line of defense against the Higher Self. If the rest of the Conscious Self is not willing to listen to the old ways, the Ego can try to reconnect us through stored conflicts in the body. Take notice if this is happening and be vigilant. And learn to work on all levels. Also past karmas from other lifetimes or from previous events in this lifetime may come due. This is why it's important to release fear and all lower negative attractor patterns in the body which can be pathways to such upheavals. Become present to transmuting old limiting karmas while not creating new ones. Clearing old karmas so you can continue operating from the ego and primal self can only run you into a new quagmire or unforeseen challenges via the subconscious mind.

As you wake up and start creating changes in your mind, subconscious mind, emotions, patterns, nervous system and DNA you can move forward in a new and unfettered life. The unravelling of the Old Self and the installation of the New Self may have a grace period of varied experiences. Remain present, remain calm and remain vigilant and your prayers, intentions, thoughts, words and actions will be answered by your reality and the Universe.

As you are changing you are changing your Akashic Records in new ways. You are getting the lessons that you may have laid out for yourself to learn through you Book of Life in a new way or even ahead of schedule. Does a person on the Earth plane incarnate in human form have to be working with the Akashic Records for any of this to occur? No. But even they are connected to their Books of Life in the Akashic Records and doing the same work you may be doing in their own way. It can be more creatively synergistic for you when you are working with your Book of Life in the Akashic Records.

Change can come quickly for you when you are working on changing yourself from multiple angles. When you become aware of such influences as ancestral DNA imprints, past life karmas, present life karmas, ego, primal self, old looping subconscious patterns, limiting beliefs or uncertainty you start to look for ways to clear the playing field and re-invent the game from a new perspective. Prayer, Meditation, Intention, Forgiveness, Release, Acceptance, Gratitude, Courage, Taking Action, Healing, Loving, Learning, Sharing, Listening and Teaching are sure fire ways to create lasting changes that will uplift you for generations to come.

Working with the Akashic Records is meant not only to be a psychic or intuitive based tool for finding messages for the past, present and future. If this is all you are doing then your ego is still trying to control your spiritual path and existence in order to insure it's own survival. Now, it is not wrong to do that and you should be doing that. If you are putting to work for yourself other forms of self help in the here and now such as healthy diet, exercise, service, giving, gratitude and forgiveness, meditation or starting to move towards any of these, then you are working smart and in a way the Akashic Records can best benefit your growing and improving life's journey as you level up.

We are here to learn, love and create. This is why in the Akashic workshops, classes, books and my personal sessions we work not only with the Akashic information, but also many ways to energize, guide, heal and transform your life. When you are in the presence of the Akashic Records and definitely through my connection with it, you are in the midst of Soul activation energy which connects you more so with your Soul and your Book of Life in the Records so transformations can start taking place in your life. Old blocks find new ways of being released as the Soul has the God Force Light of Creation radiating from it now more into your physical body, energy fields and your life.

It is a really an amazing journey to be able to experience working within the Akashic Records. So much so that as I found it or it found me decades ago, I just had an inner knowing that there was something very special about it. Self knowing through other lifetimes becomes second nature as you absorb the wisdom of just being with those experiences. It shows you that you are more than just a human being having an Earthly experience. You are a Soul.

That purpose has different variations along a theme. To learn, observe, radiate Unconditional Love and to be Creative are central components to all Souls. This gives us the insight that we are here not only to love but to create or manifest. Different Souls do it in different ways through service, leadership, healing, science, the arts, music and so many other ways. We often get lost to our Soul's Purpose. How many times I've sat in front of someone and given them their Souls's Purpose along with past life info, karmic info, future info, healing or creative venture possibilities only to have them say 'I still don't know my Soul's purpose.'

This is a great example of the Ego controlling a person's spiritual journey through selective hearing and thought processing. Others have gotten the information and all of a sudden their eyes light up because they see the correlation between their greater purpose and what they are doing in their life. If you are not getting the message on this level, not to worry, you are not being locked out of heaven. Again your ego is resisting the acknowledgement of connecting the dots because it's focused on greater than, less than, uncertainty and I'm not enough. Life gets better and better as a series of manifestations through the Universal as a result of evolution, self knowing, love and a creative approach to life. Listen and you will hear, open your eyes and you will see, knock and the door will be opened.

So continue your voyage of Self Study and employ the advantage of connecting with your Book of Life for the purpose of Self Knowing. This helps you to complete the full circle of the Soul's journey here to the Earth plane in a physical body with a complex energy field system, a chakra system, a brilliant mind, a loving heart and a Soul that knows more than you think. Born into a world where you can have, do or be anything that you choose to create. Where you can learn to use your creative power, intention and express through love. A world that is part of the Universe from which you came. Worlds upon worlds are out there, each with their own Akashic Records system for the visiting Souls to these worlds.

So many have exclaimed 'Oh this is my last life here.'

Could it be that they really know this? Or are they are just ready to bow out based on all of life's karmic and egoic challenges? Are they really doing the work to make it their last life? Or are they metaphysical dabblers that heard the message once and had already decided that life here was too much? They may be right, it might not be their bag. But while you're here, make the most of every moment in the best way that you can. Some Souls do know it's their last life here. Be grateful for every moment here and forgive yourself and others...for *everything*, not just some things. Remember you're living in a 'hall of mirrors' and the reality Universe is watching and listening. To harbor resentment might bring another unpleasant situation to you of different magnitude to give you another opportunity to release the pattern.

The more you do the inner work the clearer you will be at looking into your Records. They are watching and they know when a person is moving towards really wanting to know or understand their self on a deeper level. Often times this is when someone is guided to work with myself, another Akashic practitioner or another teacher. You may go through multiple psychic readings when you are actually looking for deeper information. Akashic information is about the cause, psychic information is about the effect. Akashic has the ability to go deeper, though, I often find that a blend of Akashic and psychic information comes through to bring the individual just what they need. See you in the Records!

Here are a few take aways from this book that we have investigated and focused on through the exercises and chapters while intending the best version of life through the Soul for you:

Time has consciousness and it does not only operate in past,present and future moments we associate with through our trained group mind. Time is just as complete in it's oneness as all things in the Universe being connected. So healing something now affects past and future. Manifesting something now recreates through Eternal Now space time. ie. Time is a ball of yarn with overlapping and intersecting points.

Your Soul is living and operating in that Eternal Now time space as you are still operating here from past, present and future moments. Practicing a blend of Eternal Now while still using linear time to function in the world will help you to understand and move old blocks out of the way to more of a non-judgmental version of your life.

By clearing the body tissue and the nervous system you will find it easier or more effortless to make changes occur whether minor or profound. You will find it easier to hold a focus and feeling of what you are intending whether it is around healing or creating new things. You will genuinely create from within rather than trying to endure what you have created unconsciously in the external.

Life will become easier as you work out the karmas of your past moments and past lives. You may be able to work out these karmas simply by acknowledging them and understanding them which can release them. Working with your Book of Life can help you release karma. By visiting the Akashic Records and activating your Soul energy karmic clearing naturally starts.

As you learn, acknowledge and experience new ways of being through being consistent with focusing your study, healing and intention in the right direction, life will clean itself up for you and move you forward into the best intended version that you can be.

This Book holds the Healing, Clearing and Manifesting abilities of the Soul energy. And by participating in the reading of this book you are naturally connecting into the Akashic Records and your own Book of Life. This is your time, so make the most of it! And when you're ready, your Book of Life will consciously be ready to meet you there within the Great Library Hall of the Akashic Records.

GLOSSARY

Akasha - A sanskrit term meaning sky, space or ether.

Akashic Records - A great way station for bringing Souls into the Earth plane for incarnation. Also a library used by God and our Souls to determine the advancement, healing and karma of the lifetimes lived. A living energetic library of every past, present, future moments of every person, place or thing, animate and inanimate.

Akashic Vision - The energetic inner visionary pictures, visions and movies that are experienced during Akashic Records, readings, practices and journeys. Akashic Vision can be dream like bringing intermittent dream like images to very lucid soul frequency movies which are interactive and can stream energy into a person through the interaction with the experience.

Alternate Realities - A universe hypothesized in some cosmological models to exist along with our own universe, possibly obeying different physical laws and having the potential for the transfer of information between universes.

Ascension - The rise of the '*Christ-consciousness*' in mankind to the point that the individual is beyond the powers of reincarnation and karma. The word '*resurrection*' as found in the New Testament is best translated as 'ascension'. After millennia of reincarnation, the soul finally gets off the wheel of karma in '*ascension*'. The transformation of a person's current ego with the light body into an ascended state of existence being able to move between dimensions including the earth plane.

Asana - A body position, typically associated with the practice of Yoga, originally identified as a mastery of sitting still. In the context of Yoga practice, asana refers to two things: the place where a practitioner (or yogin, in general usage), yogi (male), or yogini (female) sits and the manner (posture) in which he / she sits. In the Yoga sutras, Patanjali suggests that asana is "to be seated in a position that is firm, but relaxed" for extended, or timeless periods. As a repertoire of postures were promoted to exercise the body-mind over the centuries to the present day, when yoga is sought as a primarily physical exercise form, modern usage has come to include variations from lying on the back and standing on the head, to a variety of other positions. [5] However, in the Yoga sutras, Patanjali mentions the execution of sitting with a steadfast mind for extended periods as the third of the eight limbs of Classical or Raja yoga,but does not reference standing postures or kriyās. Yoga practitioners (even those who are adepts at various complex postures) who seek the "simple" practice of chair-less sitting generally find it impossible or surprisingly grueling to sit still for the traditional minimum of one hour (as still practiced in eastern Vipassana), some of them then dedicating their practice to sitting asana and the sensations and mind-states that arise and evaporate in extended sits.

Ascended Masters - Spiritually enlightened beings who in past incarnations were ordinary humans, but who have undergone a series of spiritual transformations. The Ascended Master Teachings refer to the Sixth Initiation as Ascension. According to the Ascended Master Teachings, a 'Master' (or 'Spiritual Master') is a human being who has taken the Fifth Initiation and is thereby capable of dwelling on the 5th dimension. An 'Ascended Master' is a human being who has taken the Sixth Initiation and is thereby capable of dwelling on the 6th dimension. An 'Ascended Master' is a human being who has regained full union with his 'I AM Presence.' When a human being has regained full union with his 'I AM Presence,' that state of full union is referred to as 'Ascension' Technically, a human being 'ascends' when he takes the Sixth Initiation.

Astral Plane - The 5th Dimension. The 'other-wordly' plane of existence closest to the 3D physical reality. This dimension acts as a go-tween for all of the many souls, spirits and entities interacting with humans or going on to experience other worlds. Also associated with psychic activity.

Awareness - having knowledge or perception of a situation or fact. Concerned and well-informed about a particular situation or development.

Ayurvedic - the traditional Hindu system of medicine, which is based on the idea of balance in bodily systems and uses diet, herbal treatment, and yogic breathing. Based around the principles of the elements: Earth, Water, Fire, Wood, Air. From Sanskrit āyus 'life' + veda 'science.'

Bhodisattva - in Buddhism one who has attained prajna, or Enlightenment, but who postpones Nirvana in order to help others to attain Enlightenment.

Brain Waves:

Beta waves (38 - 15 Hz) - Are the brainwaves of our "normal" waking consciousness, of our outward attention, of logical, conscious and analytical thinking. High frequency beta ("splayed beta") is seen with restlessness, stress, anxiety, panic or while our inner critic or commentator is active. Splayed beta can be differentiated from the low frequency beta of the awakened mind, when thinking feels clear, alert, creative and to the point.

Alpha brainwaves (14 - 8 Hz) are seen when we are in a relaxed state, daydreaming or visualizing ("sensualising" seems to be more appropriate as imagination in all senses - hearing, kinesthetic, smell, taste etc. - stimulates alpha waves. Your visual sense may not necessarily be the strongest for you. Some people rather feel an inner knowing). We need alpha waves as the bridge to the lower frequencies of the subconscious (theta), if we want to remember the content of our dreams or our meditation, or if we want to retrieve information from our subconscious. For this reason alpha is especially important in combination with other brainwaves.

Theta brainwaves (7 - 4 Hz) Represents the subconscious. We see theta during dream sleep (REM sleep), meditation, during peak experiences and creative states. In theta we find unconscious or suppressed parts of our psyche as well as our creativity and spirituality. Theta images are usually less distinct and colorful than alpha images, sometimes of a blueish color, but they often feel more profound and meaningful. As long as we only produce theta brainwaves, their content will stay inaccessible to our waking mind. We need alpha to bridge the gap between theta and beta brainwaves to consciously experience or remember theta content.

Delta brainwaves (3 - 0.5 Hz) Are the brainwaves of the lowest frequency and represent the unconscious. If we only produce delta we will find us in dreamless deep sleep, but we also see delta in various combinations with other brainwaves. They may then represent intuition, curiosity, a kind of radar, hunches or a "feeling" for situations and other people. Delta is often seen with people who work in therapeutic environments or professions and with people who have had traumatic experiences and have developed a "radar" for difficult situations.

Gamma brainwaves (100 - 38 Hz) were detected later than the other brainwaves, less is known about them so far. They have been seen in states of peak performance (both physical and mental), high focus and concentration and during mystic and transcendental experiences. A lot of research is currently being done on gamma brainwaves in the 40 Hz range during meditation. One of the characteristics of gamma waves is a synchronisation of activity over wide areas of the brain. Gamma brainwaves are not easy to detect because of their low amplitude and can only partly be displayed on the Mind Mirror screen. Sometimes they may be seen as a narrow frequency band at 38 Hz.

Buddhism - A religion and philosophy indigenous to the Indian subcontinent and encompasses a variety of traditions, beliefs, and practices largely based on teachings attributed to Siddhartha Gautama, who is commonly known as the Buddha (meaning '*the awakened one*' in Sanskrit and Pāli). The Buddha lived and taught in the eastern part of Indian subcontinent some time between the 6th and 4th centuries BCE. He is recognized by Buddhists as an awakened or enlightened teacher who shared his insights to help sentient beings end suffering through eliminating ignorance, craving, and hatred, by way of understanding and seeing dependent origination and no-self, and thus attain the highest happiness, nirvana. The originating teacher of Vipassana.

Bloodline - Direct line of descent; pedigree. Your ancestors, relatives and heritage.

Chakra - (in Eastern thought) each of the centers of

spiritual power in the human body, seven in number along the spine from tailbone to crown of the head. From Sanskrit chakra '*wheel* or *circle*'.

Chi (also ki,ka) The vital force believed in Taoism and other Chinese doctrines, spiritual and religious practices, thought to be inherent in all things. The unimpeded circulation of chi and a balance of its negative and positive forms in the body are held to be essential to good health in traditional Chinese medicine. Literal translation: energy. Also known as ki (*Japanese*) and ka (*Egyptian*)

Clairaudience - (*French - clair meaning 'clear' and audience meaning "hearing"*) A form of clairvoyant extra-sensory perception in which a person acquires information by paranormal auditory means. The ability to hear in a paranormal manner. May refer not to actual perception of sound, but may instead indicate impressions of the "inner mental ear" or *"the 3rd ear"*. Perception of sounds, voices, tones, or noises which are not normally audible. A clairaudient (person) might hear the voices or thoughts of the spirits, messages from God, angels, masters, gurus, or persons who are discarnate, or on the other side.

Clairsentience - (*French - clair meaning 'clear'and 'sentience' is derived from the Latin sentire,'to feel'*) A form of clairvoyant extra-sensory perception wherein a person acquires psychic knowledge by feeling. One of the six special human functions mentioned in Buddhism. Refers to a person who can feel the vibration of other people. There are many different degrees of clairsentience ranging from the perception of diseases or sickness of other people *(aka medical intuitive)* to the thoughts or emotions of other people. It differs from third eye activity in that instead of vivid pictures in the mind, a very vivid feeling can form.

Psychometry is related to clairsentience.*(Psyche and metric, which means 'soul-measuring').*

Clairvoyance - (*French - clair meaning 'clear' and voyance meaning 'vision'*) is used to refer to the ability to gain information about an object, person, location or physical event through means other than the known human senses, a form of extra-sensory perception (E.S.P.). A person said to have the ability of clairvoyance is referred to as a clairvoyant -*'one who sees clearly'*.

Consciousness - The quality or state of being aware especially within ones self. A sense of one's personal or collective identity, including the attitudes, beliefs, and sensitivities held by or considered characteristic of an individual or group. Totality of conscious states of an individual.

De ja vu - *(French - literally 'already seen')* is the impression that one has already witnessed or experienced a current situation, even though the exact circumstances of the prior encounter are unclear and were perhaps imagined. The term was coined by a French psychic researcher, Émile Boirac (1851–1917) in his book L'Avenir des sciences psychiques ('The Future of Psychic Sciences')

Deities - A being, natural, supernatural or preternatural, with superhuman powers or qualities, and who may be thought of as holy, divine, or sacred. Believers consider that they can communicate with the deity, who can respond supernaturally to their entreaties. Deities are depicted in a variety of forms, but are also frequently expressed as having human form. Deities are often thought to be immortal, and are commonly assumed to have personalities and to possess consciousness, intellects, desires, and emotions comparable but usually superior to those of humans.

Disincarnate - Having no material body or form. Souls who are between lifetimes of incarnation. unchanging, eternal, and mysterious Ein Sof (no end)

Divination - (from Latin divinare '*to foresee, to be inspired by a god*', related to divinus, divine) is the attempt to gain insight into a question or situation by way of an occultic standardized process or ritual. Used in various forms for thousands of years, diviners ascertain their interpretations of how a querent should proceed by reading signs, events, or omens, or through alleged contact with a supernatural agency. Divination can be seen as a systematic method with which to organize what appear to be disjointed, random facets of existence such that they provide insight into a problem at hand. If a distinction is to be made between divination and fortune-telling, divination has a formal or ritual and often social character, usually in a religious context, as seen in traditional African medicine; while fortune-telling is a more everyday practice for personal purposes. Particular divination methods vary by culture and religion.

DNA - *Deoxyribonucleic Acid* is a nucleic acid containing the genetic instructions used in the development and functioning of all known living organisms (with the exception of RNA viruses). The DNA segments carrying this genetic information are called genes. Likewise, other DNA sequences have structural purposes, or are involved in regulating the use of this genetic information. Along with RNA and proteins, DNA is one of the three major macromolecules that are essential for all known forms of life.

Ego - A person's sense of self-esteem or self-importance: a boost to my ego. Psychoanalysis the part of the mind that mediates between the conscious and the unconscious and is responsible for reality testing and a sense of personal identity.

Elohim - A group of special or higher order of angels who are in charge of utilizing the creation forces of the Light of Creator God. Known also to to work with the Council of Twelve and as guardians of the Akashic Records.

Empathy or Empathic ability - The action of understanding, being aware of, being sensitive to, and vicariously experiencing the feelings, thoughts, and experience of another in either the past or present time line without previously having those feelings, thoughts, and experience fully communicated in an objectively explicit manner about the person, feelings, or situation. Someone known to use empathic abilities may be referred to as an 'empath'.

Enlightenment - Spiritual revelation or deep insight into the meaning and purpose of all things, communication with or understanding of the mind of God, profound spiritual understanding or a fundamentally changed consciousness whereby everything is perceived as a unity. Freedom from desire and other worldly passions. For Hindus, as for Buddhists and Jains, enlightenment ends the cycle of reincarnation. Souls are held to enter many different bodies through the course of their existence. In each of the lives they lead they develop spiritually. Enlightenment is a state of freedom from the ignorance that causes suffering.

E.S.P. (Extra Sensory Perception) - Psychic abilities including but not limited to: mind reading, future sight, second sight, de ja vu, remote viewing, etc. Involving the acquisition or effect of past, present or future information that cannot be deduced from presently available and normally acquired sense-based information or laws of physics and/or nature.

Etherioplasma - the term is a combination of ether (the invisible energy that carries the human spirit that survives each death to form a new physical body) and the plasma energy that transfers the subtle electromagnetic forces from the soul to the human form that the human spirit use to help animate the human form. Etherio – plasmic. The actual measureable yet invisible energy substance of the Soul.

Familiars - Familiar spirits *(sometimes referred to simply as 'familiars')* were supernatural entities believed to assist witches and cunning folk in their practice of magic. According to folklore, they would appear in numerous guises, often as an animal, but also at times as a human or humanoid figure, and were described as *"clearly defined, three-dimensional... forms, vivid with color and animated with movement and sound"* by those alleging to have come into contact with them, unlike later descriptions of ghosts with their "smoky, undefined forms. Also referred to as Animal spirits who work with Shamans of different cultures as well as in Native American spiritualism.

Flower of Life - The Flower of Life is a name for a geometrical figure composed of multiple evenly-spaced, overlapping circles. This figure, used as a decorative motif since ancient times, forms a flower-like pattern with the symmetrical structure of a hexagon. A "Flower of Life" figure consists of seven or more overlapping circles, in which the center of each circle is on the circumference of up to six surrounding circles of the same diameter. However, the surrounding circles need not be clearly or completely drawn; in fact, some ancient symbols that are claimed as examples of the Flower of Life contain only a single circle or hexagon.

Grace - It has been defined as the divine influence which operates in humans to regenerate and sanctify, to inspire virtuous impulses, and to impart strength to endure trial and resist temptation and as an individual virtue or excellence of divine origin. It can also be perceived as divine intervention, blessings that were not asked for or even undeserved blessings.

Gregorian Time - The system of dates used by most of the world. The Gregorian calendar was proposed by the Calabrian doctor Aloysius Lilius and was decreed by, and named after, Pope Gregory XIII on 1582-02-24. It corrected the Julian calendar whose years were slightly longer than the solar year. It also replaced the lunar calendar which was also out of time with the seasons. The correction was achieved by skipping several days as a one-off re-synchronization and then dropping three leap days every 400 hundred years. In the revised system, leap years are all years divisible by 4 but excluding those divisible by 100 but including those divisible by 400. This gives a mean calendar year of 365.2425 days = 52.1775 weeks = 8,765.82 hours = 525,949.2 minutes = 31,556,952 seconds. Leap seconds are occasionally added to this to correct for irregularities in the Earth's rotation.

Guardian Angels - An angel assigned to protect and guide a particular person or group. The appearance of guardian angels can be traced throughout all antiquity. The concept of tutelary angels and their hierarchy was extensively developed in Christianity in the 5th century by Pseudo - Dionysius the Areopagite.

Guru - a Sanskrit term for '*teacher*' or '*master*', especially in Eastern or Indian religions. The Hindu guru - shishya tradition is the oral tradition or religious doctrine transmitted from teacher to student. A teacher and guide in spiritual and philosophical matters. A trusted counselor and adviser; a mentor. A personal spiritual teacher.

Hatha Yoga - Traditional hatha yoga is a holistic yogic path, including disciplines, postures (asana), purification procedures (shatkriya), gestures (mudra), breathing (pranayama), and meditation. The hatha yoga predominantly practiced in the West consists of mostly asanas understood as physical exercises. It is also recognized as a stress-reducing practice.

Hinduism - The predominant religion of the Indian subcontinent, and one of its indigenous religions. Among other practices and philosophies, Hinduism includes a wide spectrum of laws and prescriptions of '*daily morality*' based on the notion of karma, dharma, and societal norms. Hinduism is a conglomeration of distinct intellectual or philosophical points of view, rather than a rigid common set of beliefs. Hinduism is formed of diverse traditions and has no single founder. Among its direct roots is the historical Vedic religion of Iron Age India and, as such, Hinduism is often called the '*oldest living major religion*' in the world.

I Am That I Am or Holy I Am Presence - The Self begins with that which is the permanent atom of being and the cause out of which the effect proceeds. We call this cause the I AM THAT I AM, the Presence of the I AM, or the I AM Presence.

I find that God by any name can be reduced to this sense of the eternal Presence. It defines being, and I see it as a sphere of intense light that marks the point of my origin. It is the permanent part of me, of which I am very aware, and the point to which I will return at the conclusion of this life.

Incarnation - literally means embodied in flesh or taking on flesh. It refers to the conception and birth of a sentient creature (generally a human) who is the material manifestation of a soul, entity, god or God, whose original nature is non-physical. In its religious context the word is used to mean the descent from Heaven of a god, or divine being in human/animal form on Earth.

Introspection - (or internal perception) Is the self-examination of one's conscious thoughts and feelings. The process of introspection relies exclusively on the purposeful and rational self-observation of one's mental, physical, emotional and spiritual state; however, introspection is sometimes referenced in a spiritual context as the examination of one's soul. Introspection is the act of human self-reflection, and opposite to external observation.

Intuition - The ability to understand something immediately, without the need for conscious reasoning. The ability to acquire knowledge without the use of reason. The act or faculty of knowing or sensing without the use of rational processes; immediate cognition The word 'intuition' comes from the Latin word *'intueri'*, which is often roughly translated as meaning *'to look inside"* or *'to contemplate'*. Intuition may provide us with information that we cannot justify by ordinary means.

Karma - *Pali: kamma* means action, work or deed; it also refers to the principle of causality where intent and actions of an individual influence the future of that individual. Good intent and good deed contribute to good karma and future happiness, while bad intent and bad deed contribute to bad karma and future suffering. Karma is closely associated with the idea of rebirth in some schools of Asian religions. In these schools, karma in the present affects one's future in the current life, as well as the nature and quality of future lives - or, one's saṃsāra. In Hinduism & Buddhism, The total effect of a person's actions and conduct during the successive phases of the person's existence, regarded as determining the person's destiny. Once karmas are experienced and repaid, one may leave the cycle of reincarnation. Related to cause and effect and the law of attraction.

Kriya Yoga - Described by its practitioners as the ancient Yoga system revived in modern times by Mahavatar Babaji through his disciple Lahiri Mahasaya, circa 1861, and brought into popular awareness through Paramahansa Yogananda's book '*Autobiography of a Yogi*'. The system consists of a number of levels of Pranayama based on techniques that are intended to rapidly accelerate spiritual development and engender a profound state of tranquility and God-communion The term Kriya Yoga was developed in North India from an ancient tradition. The root of the Sanskrit word literally means '*to do*' and a true 'Kriya' technique always involves work with the body and the mind simultaneously. Kriya is a form of meditation involving Tantric Shakti flow of subtle energy within the practitioner's mind & body.

Kundalini - Described as a sleeping, dormant potential force in the human organism. It is one of the components of an esoteric description of the 'subtle body', which consists of nadis (energy channels), chakras (psychic centres), prana (subtle energy), and bindu (drops of essence). Kundalini is described as being coiled up at the base of the spine, usually within muladhara (base) chakra. The image given is that of a serpent coiled three and a half times around a smokey grey lingam. Each coil is said to represent one of the three gunas, with the half coil signifying transcendence. Through meditation, and various esoteric practices, such as Kundalini Yoga, Sahaja Yoga, and Kriya Yoga, the kundalini is awakened, and can rise up through the central nadi, called sushumna, that rises up inside or alongside the spine in the spinal fluid. The progress of kundalini through the different chakras leads to different levels of awakening and mystical experience, until the kundalini finally reaches the top of the head, Sahasrara chakra, producing an extremely profound mystical experience that is said to be indescribable.

Knowing Way of Truth and Light - The original Christ teachings which came from Atlantis. Also known as the Way Teachings, The Way, The Knowing Way.

Love - An emotion of a strong affection and personal attachment. Love is also a virtue representing all of human kindness, compassion, and affection - The unselfish loyal and benevolent concern for the good of another. Love may describe actions towards others or oneself based on compassion or affection. Love refers to a variety of different feelings, states, and attitudes, ranging from pleasure (*"I loved that meal"*) to interpersonal attraction (*"I love my partner"*). 'Love' may refer specifically to the passionate desire and intimacy of romantic love, to the sexual love of eros, to the emotional closeness of family love, to the platonic love that defines friendship, or to the profound oneness or devotion of religious love, or to a concept of love that encompasses all of those feelings. This diversity of uses and meanings, combined with the complexity of the feelings involved, makes love unusually difficult to consistently define, compared to other emotional states. Love in its various forms acts as a major facilitator of interpersonal relationships and, owing to its central psychological importance, is one of the most common themes in the creative arts.

Lymph (glands) - Physiology a colorless fluid containing white blood cells, that bathes the tissues and drains through the lymphatic system into the bloodstream. Fluid exuding from a sore or inflamed tissue. Literary pure water.

Mala - A string of beads worn around the neck or wrist, traditionally of 108 bead count, to be used for prayers, mantras and chants. From the ancient Eastern traditions, and can also refer to a catholic rosary.

Mandala - A Sanskrit word meaning *'circle.'* In the Buddhist and Hindu religious traditions sacred art often takes a mandala form. The basic form of most Hindu and Buddhist mandalas is a square with four gates containing a circle with a center point. Each gate is in the shape of a T. Mandalas often exhibit radial balance. These mandalas, concentric diagrams, have spiritual and ritual significance in both Buddhism and Hinduism.

The term is of Hindu origin and appears in the Rig Veda as the name of the sections of the work, but is also used in other Indian religions, particularly Buddhism. In the Tibetan branch of Vajrayana Buddhism, mandalas have been developed into sand painting. They are also a key part of anuttara yoga tantra meditation practices. In various spiritual traditions, mandalas may be employed for focusing attention of aspirants and adepts, as a spiritual teaching tool, for establishing a sacred space, and as an aid to meditation and trance induction. It's symbolic nature can help one "to access progressively deeper levels of the unconscious, ultimately assisting the meditator to experience a mystical sense of oneness with the ultimate unity from which the cosmos in all its manifold forms arises." The psychoanalyst Carl Jung saw the mandala as *"a representation of the unconscious self,"* and believed his paintings of mandalas enabled him to identify emotional disorders and work towards wholeness in personality.

Mantras - a sound, seed sound, syllable, word, or group of words that is considered capable of creating (spiritual) transformation when used as a written or chanted prayer. A sacred verbal formula repeated in prayer, meditation, or incantation, such as an invocation of God, a magic spell, or a syllable or portion of scripture containing mystical potentialities. Chanted individually and in groups both silently and aloud traditionally 108 times and with the use of a mala or prayer beads.

Mediumship - Involves a cooperating effort between a person on the Earth plane (the medium or channel) and a person in Spirit (the communicator).

Mudras - In Sanskrit: '*seal*', '*mark*', or '*gesture*'; Tibetan, *chakgya* is a symbolic or ritual gesture in Hinduism and Buddhism. While some mudrās involve the entire body, most are performed with the hands and fingers. A mudrā is a spiritual gesture and an energetic seal of authenticity employed in the iconography and spiritual practice of Indian religions and traditions of Dharma and Taoism. One hundred and eight mudrasare used in regular Tantric rituals. In yoga, mudrās are used in conjunction with pranayama (yogic breathing exercises), generally while seated in Padmasana, Sukhasana or Vajrasana pose, to stimulate different parts of the body involved with breathing and to affect the flow of prana in the body.

Meditation - A practice in which an individual trains

the mind or induces a mode of consciousness, either to realize some benefit or as an end in itself. The term meditation refers to a broad variety of practices (much like the term sports) that includes techniques designed to promote relaxation, build internal energy or life force (qi, ki, prana, etc.) and develop compassion, love, patience, generosity and forgiveness. A particularly ambitious form of meditation aims at effortlessly sustained single-pointed concentration single-pointed analysis, meant to enable its practitioner to enjoy an indestructible sense of well-being while engaging in any life activity.

Metatron's Cube - The name of Metatron's Cube makes reference to Metatron, an angel mentioned in apocryphal texts including the Second Book of Enoch and the Book of the Palaces.These texts rank Metatron second only to the Abrahamic God in the hierarchy of spiritual beings. The derivation of Metatron's cube from the tree of life, which the Talmud clearly states was excluded from human experience during the exile from Eden.

Metatron's cube contains every shape that exists in the universe God has created, and those shapes are the building blocks of all physical matter, which are known as Platonic solids (because the philosopher Plato linked them to the spirit world of heaven and the physical elements on Earth).

The pattern delineated by many of the lines can be created by orthographic projections of the first three Platonic solids. Specifically, the line pattern includes projections of a double tetrahedron (aka stellated octahedron), a cube within a cube (a three-dimensional projection of a tesseract), and an octahedron.

Monad - Was a term for Divinity or the first being, or the totality of all beings. Monad being the source or the One meaning without division.

Mysticism - from the Greek, mystikos, meaning 'an initiate') is the knowledge of, and especially the personal experience of, states of consciousness, or levels of being, or aspects of reality, beyond normal human perception, including experience of and even communion with a supreme being.

Neocortex - The neocortex consists of the grey matter, or neuronal cell bodies and unmyelinated fibers, surrounding the deeper white matter (myelinated axons) in the cerebrum. The neocortex is smooth in rodents and other small mammals, whereas in primates and other larger mammals it has deep grooves (sulci) and wrinkles (gyri). These folds allow the surface area of the neocortex to increase far beyond what could otherwise be fit in the same size skull. All human brains have the same overall pattern of main gyri and sulci, although they differ in detail from one person to another. The mechanism by which the gyri form during embryogenesis is not entirely clear. However, it may be due to differences in cellular proliferation rates in different areas of the cortex early in embryonic development.

Om or Aum - is a mystical sound of Sanskrit origin, sacred and important in various Dharmic religions such as Hinduism, Buddhism, and Jainism. It is placed at the beginning of most Hindu texts as a sacred incantation to be intoned at the beginning and end of a reading of the Vedas or prior to any prayer or mantra. It is used at the end of the invocation to the god being sacrificed to (anuvakya) as an invitation to and for that God to partake of the sacrifice. The Māndukya Upanishad is entirely devoted to the explanation of the syllable. The syllable consists of three phonemes, a (Vaishvanara), u (Hiranyagarbha), and m(Ishvara), which symbolize the beginning, duration, and dissolution of the universe and the associated gods Brahma, Vishnu, and Shiva, respectively. The name omkara is taken as a name of God in the Hindu revivalist Arya Samaj and can be translated as *"I Am Existence"*. Also refered to as the primordial vibrational tone behind all realities and dimensions. Monks and Siddhas have reported to hear this tone vibrating in the deepest levels of trance and meditation.

Om Mani Padme Hum - *"The jewel of consciousness is in the lotus of my heart"* and *"I bow to the light within"* are meditative translations to focus upon while using this mantra. Om coincides with the 3rd eye and forehead, Mani - back of the head, Padme - heart and Hum - throat. Visualizing a ring of God's Divine light through these centers while chanting this mantra clears away negativity, pain, fear and stress and brings compassion and healing to the heart, mind and body, while opening and blending what the higher teachings refer to as the Heart/Mind. Tibetan Buddhists believe that saying this mantra (prayer), Om Mani Padme Hum, out loud or silently to oneself, invokes the powerful benevolent attention and blessings of Chenrezig, the embodiment of compassion. Viewing the written form of the mantra is said to have the same effect. It is said that all the teachings of the Buddha are contained in this mantra.

According to the Dali Lama: "It is very good to recite the mantra ***Om Mani Padme Hum***, but while you are doing it, you should be thinking on it's meaning, for the meaning of the six syllables is great and vast... The first, *'Om'* symbolizes the practitioner's impure body, speech, and mind; it also symbolizes the pure exalted body, speech, and mind of a Buddha. The path is indicated by the next four syllables. '*Mani*', meaning jewel, symbolizes the factors of method: (the) altruistic intention to become enlightened, compassion, and love. The two syllables, '*Padme*', meaning lotus, symbolize wisdom. Purity must be achieved by an indivisible unity of method and wisdom, symbolized by the final syllable '*Hum*', which indicates indivisibility. Thus the six syllables, '*Om Mani Padme Hum*', mean that in dependence on the practice of a path which is an indivisible union of method and wisdom, you can transform your impure body, speech, and mind into the pure exalted body, speech, and mind of a Buddha"

Omnipresence - The property of being present everywhere. This characteristic is most commonly used in a religious context, as most doctrines bestow the trait of omnipresence onto a superior, a deity or God. This also identifies the universe and divinity; in divine omnipresence, the divine and universe are separate, but the divine is present everywhere.

Hinduism, and other religions that derive from it, incorporate the theory of transcendent and immanent omnipresence which is the traditional meaning of the word, Brahman. This theory defines a universal and fundamental substance, which is the source of all physical existence. Divine omnipresence is thus one of the divine attributes.

Past Lives - According to the study of reincarnation, the lives that our soul has previously lived or participated in. Past lives can be incarnations within the same family or within different races and regions of the world. These past lives are often discoverable through dreams, past life regression, Shamanic Journeys and Journeys into the Akashic Records and a person's Book of Lifetimes.

Pan Fairy Realm - The 2nd Dimensional beings that often interact with nature. Fairies, gnomes, earth and nature devas that assist mother earth in tending her planetary garden. Intermittently these being can find their way into human incarnation by way of evolution. Often these folks can feel a deep bond with mother nature without knowing why. Some incarnate knowing where they came from.

Pineal Gland - (or the '*third eye*') is a small endocrine gland in the brain. It produces the serotonin derivative melatonin, a hormone that affects the modulation of wake/sleep patterns and seasonal functions. Its shape resembles a tiny pine cone (hence its name), and it is located near the centre of the brain, between the two hemispheres, tucked in a groove where the two rounded thalamic bodies join. The Pineal Gland has for long been associated with Esoteric Knowledge surrounding the spiritual, metaphysical aspects of consciousness and the self. René Descartes, who dedicated much time to the study of the pineal gland, called it the '*Seat of the Soul*'. He believed that it was the point of connection between the intellect and the body. Directly behind the root of the nose (3rd eye chakra) floating in a small lake of cerebrospinal fluid. It is our body's biological clock. The pineal gland has been supplied with the best blood, oxygen and nutrient mix available other than that received by our kidneys. It acts as a receiving mechanism capable of monitoring electro-magnetic fields and helping align bodies in space. With its central hormone, Melatonin, the pineal not only regulates sleep/wake cycles and the aging process, but also appears to act as the Mistress Gland (sofia)* orchestrating the body's entire endocrine system and thus, energetically speaking, the chakra system. It is also responsible for shamanic states, visions, kundalini awakening e.t.c.

Pituitary Gland - An endocrine gland about the size of a pea and weighing 0.5 grams (0.018 oz) in humans. It is not a part of the brain. It is a protrusion off the bottom of the hypothalamus at the base of the brain, and rests in a small, bony cavity (sella turcica) covered by a dural fold (diaphragma sellae). The pituitary is functionally connected to the hypothalamus by the median eminence via a small tube called the infundibular stem (Pituitary stalk). The pituitary fossa, in which the pituitary gland sits, is situated in the sphenoid bone in the middle cranial fossa at the base of the brain. The pituitary gland secretes nine hormones that regulate homeostasis. The Pituitary Gland is known as the Master Gland of the Endocrine System. It's secretions regulate all the other Endocrine Glands. This gland represents one's ability to coordinate the different aspects of one's Life. Problems represent difficulty doing this. The Pituitary Gland is linked to the Hypothalamus, also located in the brain, whose function is to maintain Homeostasis in the body. That is the body's tendency to return automatically to its level of highest functioning. Metaphysically, this means aligning the frequencies of the physical and energetic bodies to homeostatically return to one's highest spiritual functioning. The pituitary gland is called the *"Seat of the Mind"* with the frontal lobe regulating emotional thoughts such as poetry and music, and the anterior lobe regulating concrete thought and intellectual concepts. The pineal gland is known as the *'Seat of Illumination, Intuition and Cosmic Consciousness'*. The pineal gland is to the pituitary gland what intuition is to reason. The glandular system also coincides with the chakra system.

Prana - The Sanskrit word for *'vital life'* (from the root prā 'to fill',). It is one of the five organs of vitality or sensation, prana *"breath"*, vac *'speech'*, chakshus *'sight'*, shrotra *'hearing'*, and manas *'thought'* (nose, mouth, eyes, ears and mind). In Vedantic philosophy, prana is the notion of a vital, life-sustaining force of living beings and vital energy, comparable to the Chinese notion of Qi. Prana is a central concept in Hinduism, particularly in Ayurveda and Yoga. It flows through a network of fine subtle channels called nadis. Its most subtle material form is the breath, but it is also to be found in the blood, and its most concentrated form is semen in men and vaginal fluid in women. Prana was first expounded in the Upanishads, where it is part of the worldly, physical realm, sustaining the body and the mother of thought and thus also of the mind. Prana suffuses all living forms but is not itself the Atman or individual soul. In the Ayurveda, the Sun and sunshine are held to be a source of prana.

Pranayama - A Sanskrit word meaning 'extension of the prana or breath' or more accurately, *'extension of the life force'*. The word is composed of two Sanskrit words, Prāna, life force, or vital energy, particularly, the breath, and 'ayāma', to extend or draw out.

Psyche - The totality of the human mind, conscious, and unconscious. Psychology is the scientific or objective study of the psyche. The word has a long history of use in psychology and philosophy, dating back to ancient times, and has been one of the fundamental concepts for understanding human nature from a scientific point of view.

Psychic - Relating to or denoting faculties or phenomena that are apparently inexplicable by natural laws, especially involving telepathy or clairvoyance: psychic powers. A person appearing or considered to have powers of telepathy or clairvoyance. Of or relating to the soul or mind. From the Greek psychikos—'*of the mind, mental*', is a person who possesses an ability to perceive information hidden from the normal senses through extrasensory perception (ESP), It can also denote an ability of the mind to influence the world physically using psychokinetic powers. Elaborate systems of divination and fortune - telling date back to ancient times. Perhaps the most widely-known system of early civilization fortune - telling was astrology, where practitioners believed the relative positions of celestial bodies could lend insight into people's lives and even predict their future circumstances. Some fortune-tellers were said to be able to make predictions without the use of ritualistic objects or special, spiritual, or energy tools for diving information. More so through direct apprehension or vision of the past, present or future. These people were known as seers or prophets, and in later times as clairvoyants and psychics.

Qi Gong - a practice of aligning breath, movement, and awareness for exercise, healing, and meditation. With roots in Chinese medicine, martial arts, and philosophy, qigong is traditionally viewed as a practice to cultivate and balance qi (chi) or what has been translated as '*intrinsic life energy*'. Typically a Qigong practice involves rhythmic breathing coordinated with slow stylized repetition of fluid movement, a calm mindful state, and visualization of guiding qi through the body. Qigong is now practiced throughout China and worldwide, and is considered by some to be exercise, and by others to be a type of alternative medicine or meditative practice. From a philosophical perspective Qigong is believed to help develop human potential, allow access to higher realms of awareness, and awaken one's 'true nature'.

Quantum Physics - The study of the behavior of matter and energy at the molecular, atomic, nuclear, and even smaller microscopic levels. In the early 20th century, it was discovered that the laws that govern macroscopic objects do not function the same in such small realms. In the realm of quantum physics, observing something actually influences the physical processes taking place. Light waves act like particles and particles act like waves (called wave particle duality). Matter can go from one spot to another without moving through the intervening space (called quantum tunnelling). Information moves instantly across vast distances. In fact, in quantum mechanics we discover that the entire universe is actually a series of probabilities.

R.E.M. (raphid eye movement) - REM sleep typically occupies 20–25% of total sleep, about 90–120 minutes of a night's sleep. REM sleep is considered the deepest stage of sleep, and normally occurs close to morning. During a night of sleep, one usually experiences about four or five periods of REM sleep; they are quite short at the beginning of the night and longer toward the end. Many animals and some people tend to wake, or experience a period of very light sleep, for a short time immediately after a bout of REM. The relative amount of REM sleep varies considerably with age. A newborn baby spends more than 80% of total sleep time in REM. During REM, the activity of the brain's neurons is quite similar to that during waking hours; for this reason, the REM-sleep stage may be called paradoxical sleep.

Rosicrucians - Studies or membership within a philosophical secret society said to have been founded in late medieval Germany by Christian Rosenkreuz. It holds a doctrine or theology "built on esoteric truths of the ancient past", which, "concealed from the average man, provide insight into nature, the physical universe and the spiritual realm." Rosicrucianism is symbolized by the Rosy Cross. In the early 17th century, the manifestos caused excitement throughout Europe by declaring the existence of a secret brotherhood of alchemists and sages who were preparing to transform the arts, sciences, religion, and political and intellectual landscape of Europe. Wars of politics and religion ravaged the continent. The works were re-issued several times and followed by numerous pamphlets, favorable and otherwise. Between 1614 and 1620, about 400 manuscripts and books were published which discussed the Rosicrucian documents.

Samadhi - Described as a non-dualistic state of consciousness in which the consciousness of the experiencing subject becomes one with the experienced object, and in which the mind becomes still, one-pointed or concentrated while the person remains conscious. In Buddhism, it can also refer to an abiding in which mind becomes very still but does not merge with the object of attention, and is thus able to observe and gain insight into the changing flow of experience. In Hinduism, samādhi can also refer to videha mukti or the complete absorption of the individual consciousness in the self at the time of death, usually referred to as mahasamādhi.

Sanskrit - Classical Sanskrit is the standard register as laid out in the grammar of Pāṇini, 4th century BCE and it has significantly influenced most modern languages of the Indian subcontinent, particularly in India, Pakistan, Sri Lanka and Nepal. The pre-Classical form of Sanskrit is known as Vedic Sanskrit, with the language of the Rigveda being the oldest and most archaic stage preserved, its oldest core dating back to as early as 1500 BCE. This qualifies Rigvedic Sanskrit as one of the oldest attestations of any Indo-Iranian language, and one of the earliest attested members of the Indo-European language family.

Siddha - A Siddham in Tamil(an Indian sub-continent dialect) means '*one who is accomplished*' and refers to perfected masters who, according to Hindu belief, have transcended the ahamkara (ego or I-maker), have subdued their minds to be subservient to their Awareness, and have transformed their bodies (composed mainly of dense Rajo-tama gunas) into a different kind of body dominated by sattva. This is usually accomplished only by persistent meditation. Siddhas are the liberated souls. They have completely ended the cycle of birth and death. They have reached the ultimate state of salvation. They do not have any karmas and they do not collect any new karmas. This state of true freedom is called Moksha. They are formless and have no passions and therefore are free from all temptations. A siddha has also been defined to refer to one who has attained a siddhi.

Siddhi - The siddhis as paranormal abilities are considered emergent abilities of an individual that is on the path to siddhahood, and do not define a siddha, who is established in the Pranav or Aum – the spiritual substrate of creation. The siddhi in its pure form means "the attainment of flawless identity with Reality (Brahman); perfection of Spirit." In the Hindu philosophy of Kashmir Shaivism (Hindu tantra), siddha also refers to a Siddha Guru who can by way of Shaktipat initiate disciples into Yoga.

Soul or **Soul Group** - The incorporeal and immortal

essence of a person, living thing, or object. Souls which are immortal and capable of union with the divine belong only to human beings. 12 Souls are grouped together in a Soul Group which is governed by an Over Soul. 12 Soul Groups are governed by a Monad.

Solar Gazing - Hira Ratan Manek (HRM), among others, have proven eating any food. The method is used for curing all kinds of psychosomatic, mental and physical illnesses as well as increasing memory power and mental strength by using sunlight. One can get rid of any kind of psychological problems, and develop confidence to face any problem in life and can overcome any kind of fear including that of death within 3 months after starting to practice this method. As a result, one will be free from mental disturbances and fear, which will result in a perfect balance of mind. If one continues to apply the proper sun gazing practice for 6 months, they will be free from physical illnesses. Furthermore, after 9 months, one can eventually win a victory over hunger, which disappears by itself thereafter. This is a straight-forward yet effective method based on solar energy, which enables one to harmonize and recharge the body with life energy and also invoke the unlimited powers of the mind very easily. Additionally, it allows one to easily liberate from threefold sufferings of humanity such as mental illnesses, physical illnesses and spiritual ignorance.

Spirit Guides - A term used by the Western tradition of Spiritualist Churches, mediums, and psychics to describe an entity that remains a disincarnate spirit in order to act as a guide or protector to a living incarnated human being. Traditionally, within the spiritualist churches, spirit guides were often stereotyped ethnically, with Native Americans,

Chinese or Egyptians being popular for their perceived ancient wisdom. Other popular types of guides were saints or other enlightened individuals. The term can also refer to totems, angels, guardian angels or nature spirits.

Spirit guides are not always of human descent. Some spirit guides live as energy, in the cosmic realm, or as light beings, which are very high level spirit guides. Some spirit guides are persons who have lived many former lifetimes, paid their karmic debts, and advanced beyond a need to reincarnate. Many devotees believe that spirit guides are chosen on "the other side" by human beings who are about to incarnate and wish assistance.

Tai Chi - A type of internal Chinese martial art practiced for both its defense training and its health benefits. It is also typically practiced for a variety of other personal reasons: its hard and soft martial art technique, demonstration competitions, and longevity. As a result, a multitude of training forms exist, both traditional and modern, which correspond to those aims. Some of Tai Chi Chuan's training forms are especially known for being practiced at what most people categorize as slow movement.

Tantra or **tantric** - Defined primarily as a technique-rich style of spiritual practice, Tantra has no single coherent doctrine; rather, it developed different teachings in connection with the different religions that adopted the Tantric method. These teachings tended to support and validate the practices of Tantra, which in their classical form are more oriented to the married householder than the monastic or solitary renunciant, and thus exhibited what may be called a world-embracing rather than a world-denying character. Thus Tantra, especially in its nondual forms, rejected the renunciant values of Patañjalian yoga, offering instead a vision of the whole of reality as the self-expression of a single, free and blissful Divine Consciousness under whatever name, whether Śiva or Buddha-nature.

Since the world was viewed as real, not illusory, this doctrine was a significant innovation over and against previous Indian philosophies, which tended to picture the Divine as absolutely transcendent and/or the world as illusion. The practical consequence of this view was that not only could householders aspire to spiritual liberation in the Tantric system, they were the type of practitioner that most Tantric manuals had in mind. Furthermore, since Tantra dissolved the dichotomy of spiritual versus mundane, practitioners could entail every aspect of their daily lives into their spiritual growth process, seeking to realize the transcendent in the immanent.

Tantric spiritual practices and rituals thus aim to bring about an inner realization of the truth that "*Nothing exists that is not Divine*" (nāśivaṃ vidyate kvacit), bringing freedom from ignorance and from the cycle of suffering (saṃsāra) in the process. In fact, tantric visualizations are said to bring the meditator to the core of his humanity and oneness with transcendence. Tantric meditations do not serve the function of training or practicing extra beliefs or unnatural ways. On the contrary, the transcendence that is reached by such meditative work does not construct anything in the mind of the practitioner, but actually de-constructs all preconceived notions of the human condition. The barriers that constrict thinking to limitation, namely, cultural and linguistic frameworks are completely removed. This allows the person to experience total liberation and then unity with ultimate truth or reality.

Thymus gland - The thymus is a specialized organ of

the immune system. The thymus "*educates*" T-lymphocytes (T cells), which are critical cells of the adaptive immune system. Each T cell attacks a foreign substance which it identifies with its receptor. T cells have receptors which are generated by randomly shuffling gene segments. Each T cell attacks a different antigen. T cells that attack the body's own proteins are eliminated in the thymus. Thymic epithelial cells express major proteins from elsewhere in the body, and T cells that respond to those proteins are eliminated through programmed cell death (apoptosis). The thymus is composed of two identical lobes and is located in front of the heart and behind the sternum.

Tao - A Chinese word meaning *'way'*, *'path'*, *'route'*, or sometimes more loosely, *'doctrine'* or *'principle'*. Within the context of traditional Chinese philosophy and religion, Tao is a metaphysical concept originating with Laozi that gave rise to a religion (Wade–Giles, Tao Chiao; Pinyin, Daojiao) and philosophy (Wade–Giles, Tao chia; Pinyin, Daojia) referred to in English with the single term Taoism. The concept of Tao was later adopted in Confucianism, Chán and Zen Buddhism and more broadly throughout East Asian philosophy and religion in general. Within these contexts Tao signifies the primordial essence or fundamental nature of the universe. In the foundational text of Taoism, the Tao Te Ching, Laozi explains that Tao is not a *'name'* for a *'thing'* but the underlying natural order of the universe whose ultimate essence is difficult to circumscribe. Tao is thus *'eternally nameless'* (Dao De Jing-32. Laozi) and to be distinguished from the countless 'named' things which are considered to be its manifestations.

Violet Flame - The Violet Flame is a Divine gift and tool for everyone, given to us by Ascended Master Saint Germain.

It is a sacred fire that exists on the Higher Dimensions. People with the gift of interdimensional sight have seen it. Cameras have captured it when it was not visible to the person taking the photo. The Violet Flame is REAL and I invite you to use it to your great advantage. The Violet Flame is Spiritual Alchemy in action. Just as Alchemy is said to turn Lead into Gold, the ultimate purpose of the Violet Flame is to turn the Human into the Divine Human. Its action is to TRANSMUTE denser feelings, actions, deeds, karma, etc. into a higher vibrational frequency, which helps prepare us for our Ascension

Vipassana - In the Buddhist tradition means insight into the true nature of reality. A regular practitioner of Vipassana, is known as a Vipassi.Vipassana is one of the world's most ancient techniques of meditation, which was introduced by Gautama Buddha. It is a practice of self-transformation through self-observation and introspection to the extent that sitting with a steadfast mind becomes an active experience of change and impermanence. In the west, Vipassanā meditation is often referred to simply as *"insight meditation"*.

Wheel of Life - The bhavacakra (Sanskrit; Pali: bhavacakka; Tibetan: srid pa'i 'khor lo) is a symbolic representation of samsara (or cyclic existence) found on the outside walls of Tibetan Buddhist temples and monasteries in the Indo-Tibetan region. In the Mahayana Buddhist tradition, it is believed that the drawing was designed by the Buddha himself in order to help ordinary people understand the Buddhist teachings. The bhavacakra is popularly referred to as the wheel of life. This term is also translated as wheel of cyclic existence or wheel of becoming.

Yantra - Is the Sanskrit word for "instrument" or "machine". Much like the word "instrument" itself, it can stand for symbols, processes, automata, machinery or anything that has structure and organization, depending on context. One usage popular in the west is as symbols or geometric figures. Traditionally such symbols are used in Eastern mysticism to balance the mind or focus it on spiritual concepts. The act of wearing, depicting, enacting and / or concentrating on a yantra is held to have spiritual or astrological or magical benefits in the Tantric traditions of the Indian religions.

Akashic Records, Healing, and Personal Guidance Phone Sessions

Bill works with you in person when available or by phone to unlock your greatest potential, give life path guidance, for personal healing and manifesting your goals and dreams in life. Looking into your personal Book of Life within the Akashic Records. This enables you to see tendencies from other lifetimes as they are played out in the multi-dimensional reality that makes up who you are. Channeled information from the Records, with the aid of your spirit guides, angelics, and ascended masters or spiritual gurus, reveal information that make up who you are on the otherside, as well as in the physical. This can bring clarity to your current situation. As you state your questions, Your answers will come from the Akashic Records and will give clarity to the reasons for the immediate events and circumstances as well as how to navigate more clearly and receive info for business and personal relations, spiritual attainment, and locations, Past, Present and Future.

Book your Session: **www.billfoss.net**

"If you need help in getting to the next level on your journey,This will be beneficial for you."

AUDIO DOWNLOADS
Experience Akashic Records Access
Energy and Soul Healing Journeys
Individual Downloads and Bundles

Available at the Sites:
billfoss.net
billfoss-akashic.com

Journey to the Akashic Records Workbook
Study, Practice, and Learn New Ways of Opening Into the Akashic Records!
Available at billfoss.net or Amazon

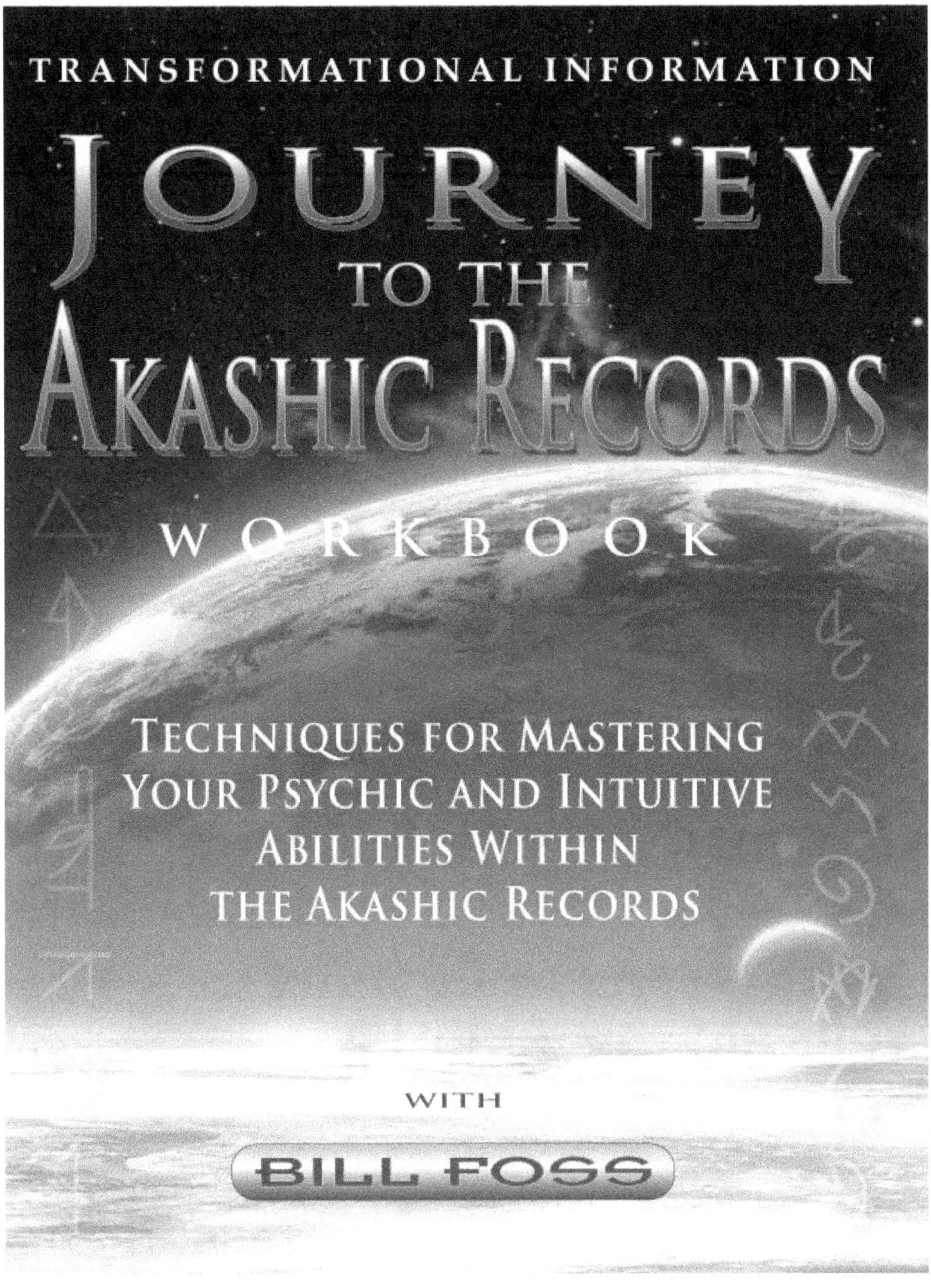

The study companion workbook for all the exercises at the workshops, presentations and classes. This workbook will help you open and expand your abilities while giving you insight to strengthen your practice or working with your intuitive and healing abilities.
Size 8.5″ x 11″ 165 Pages

NEW BEGINNINGS
VOLUME 1
HOW TO READ
INTERMEDIATE
VOLUME 2
HOW TO READ
ADVANCED ACCESS
VOLUME 3
HOW TO READ
THE
AKASHIC
RECORDS
MOVING DEEPER INTO THE RECORDS
ADVANCED DYNAMICS,
BENEFITS & TECHNIQUES
WITH
BILL FOSS
THE AKASHIC RECORDS CENTER
& INSTITUTE FOR HIGHER LEARNING
THE
AKASHIC
RECORDS
ADVANCED
WORKBOOK

Visit the Websites

billfoss.net **billfoss-akashic.com** **akashicrecordscenter.com**

The Secrets of Spiritual Success is a clear and concise road map providing an overview of understanding of 1000's of years of spiritual teachings, inquiries, and understandings. This book will help you to connect the dots more quickly in life, while adding to your studies already in motion or providing a great primer source to get started from. Whether you are a mystical student from beyond or you are new to expanded views and just looking for answers, this book was created to enliven and enrich your search, practices and studies. This book will turn on the lights for you and keep them on." Size 6" x 9" 237 Pages

Book • Workbook • Journal

Complete Study Set - Order at billfoss.net

"The Keys of Spiritual Success" Workbook, companion to "The Secrets of Spiritual Success" is a collection of creative visualization and healing techniques, energy exercises, meditations, prayers. Exercises and insights to jump start your journey into your own long awaited or continued self inquiry and realization. Use this Workbook to Journal your subtle and not so subtle experiences as you open to greater understandings, fresh ideas, and new ways of being!" Size 8.5" x 11" 165 Pages

"The "Book Of Dreams" Journal is your gateway into creativity. This is your Journal, your space to create. What will you write, sketch, plan, or invent? Take the opportunity to go within and explore the vast regions, depths, and banks of Divine Creative Potential existing within you, all around you, throughout time, space & beyond. This is your chance to write down, plan and draw out your dream and make it a reality. Use this book as you will to expand your vision. It all starts here, and it starts with you so let's begin. Size 8.5" x 11" 165 Pages

Notes:

Notes:

www.ingramcontent.com/pod-product-compliance
Lightning Source LLC
LaVergne TN
LVHW080015210926
848481LV00011B/3030